KT-460-766

Am I Thin Enough Yet?

142416

BCFTCS

£8.99

Am I Thin Enough Yet?

The Cult of Thinness and the Commercialization of Identity

Sharlene Hesse-Biber

OXFORD UNIVERSITY PRESS

New York   Oxford

B.C.F.T.C.S.

142416

Oxford University Press

Oxford   New York
Athens Auckland Bangkok Bogotá Bombay
Buenos Aires Calcutta Cape Town Dar es Salaam
Delhi Florence Hong Kong Istanbul Karachi
Kuala Lumpur Madras Madrid Melbourne
Mexico City Nairobi Paris Singapore
Taipei Tokyo Toronto Warsaw

and associated companies in
Berlin Ibadan

Copyright © 1996 by Oxford University Press, Inc.

First published by Oxford University Press, Inc., 1996

First issued as an Oxford University Press paperback, 1997

Oxford is a registered trademark of Oxford University Press

All rights reserved. No part of this publication may be reproduced,
stored in a retrieval system, or transmitted, in any form or by any means,
electronic, mechanical, photocopying, recording, or otherwise,
without the prior permission of Oxford University Press.

Library of Congress Cataloging-in-Publication Data
Hesse-Biber, Sharlene Janice.
Am I thin enough yet?: the cult of thinness
and the commercialization of identity / Sharlene Hesse-Biber.
p. cm.
Includes bibliographical references and index.
ISBN-13 978-0-19-508241-8
ISBN 0-19-508241-9
ISBN-13 978-0-19-511791-2 (Pbk.)
ISBN 0-19-511791-3 (Pbk.)
1. Body image. 2. Leanness—Psychological aspects
3. Eating disorders—Social aspects.
4. Women—Psychology. I. Title.
BF697.5.B63H47 1996
306.4—dc20   95-11812

10 9 8

Printed in the United States of America

To my strong-willed daughters,
  Sarah Alexandra Biber and Julia Ariel Biber

  and

in memory of two pioneering women,
  Carol Biber, Ph.D., and Marie Feltin, M.D.

Acknowledgments

Many in the Boston College community contributed generously to this project. My thanks go first to the students of Boston College who shared their thoughts and opinions with me in replying to the questionnaire and intensive interviews. My appreciation also goes to the faculty and staff of the Sociology Department. I would like to especially thank Dr. Jeanne Guillemin, Dr. David Karp, Dr. William Gamson, and Dr. Ritchie Lowry for reading the entire manuscript and offering insightful feedback. Brenda Pepe, Roberta Negrin, and Eunice Doherty worked hard to ensure high quality transcriptions of the interview data and provided much humor and encouragement along the way. I want to thank the Office of Research Administration of Boston College for providing research grants to support this project.

Kate Kruschwitz worked very hard, and successfully, to make my academic writing more accessible to a wider audience. She offered much advice and editorial assistance and was extremely supportive throughout the writing of this book.

Special thanks go to several graduate research assistants. In particular, James Vela-McConnell, for his insightful coding of the intensive interviews as well as providing invaluable feedback on book chapters. Thanks also to Amanda Udis-Kessler who compiled research statistics and organized some notes on the food, diet, and self-help industries. Julie Manga compiled a range of research statistics and notes on the fitness industry and assisted with reference materials.

Along the way, many others supported and assisted in this project and I am particularly grateful to Margaret Marino, John Downey, Alan Clayton-Matthews, James Meehan, Kelly Joyce, Steve Vedder, The Writer's Group, Barbara Deck and the Mother's Group of Newton, Erica Hiersteiner, Maybelle Rowntree, Lorraine Heilbrunn, Bob Ford, and Anne Tropeano.

I want to thank Gioia Stevens, social science editor at Oxford University Press, for her enthusiastic support as well as acknowledge the hard-working efforts of the editorial staff at Oxford University Press.

I am grateful to my daughters Sarah Alexandra and Julia Ariel for their patience, humor, love, kindness, and enthusiastic support for combining my work and family life. I want to thank their friends, especially the "BB& N carpool gang" for the many hours of discussion on "being a teenager." I want to also express my appreciation to the group of pre-teens from Brookine, Massachusetts, who haven't "hit the wall" and who openly and proudly shared their thoughts and advice on coming of age in the 1990s. Thanks also go to my mother, Helene Stockert, and stepfather, Harry Stockert, for their care, support, and great cooking.

I want to express my sincere appreciation to my husband, Michael, for "hanging in there" while I spent many hours on this project. His support, love, and understanding were crucial during all phases.

Contents

Am I Thin Enough Yet?

Introduction

Several years ago, the director of the Counseling Center at Boston College asked me to help find out why the Center was overwhelmed with female students reporting eating problems. The situation had been getting worse over the past few years. Numerous cases of bulimia (compulsive binge eating, often followed by self-induced vomiting) and anorexia (obsession with food, starvation dieting, and severe weight loss) appeared every week.

As the author of a previous study on female student career and lifestyle aspirations, a teacher of women's studies, and a faculty advisor, I was fascinated by the fact that eating disorders were much more common among women.[1] I wanted to understand why such problems had recently exploded. Although bulimia and anorexia are individual diagnoses, one can assume that broader factors are at work when the incidence of a disorder suddenly increases.[2] Was something going on in our society to foster such behavior?

Over the next several months I began researching the field of eating disorders. But the problem didn't strike home until one of the sophomores I was advising came into my office in tears. Janet broke into sobs and said to me, "I don't know what I am going to do. I'm too fat for the cheerleading squad."

Janet was fairly tall (5′ 8″) with a medium build. She weighed 125 pounds. She told me that when she showed up for the co-ed cheerleading tryouts, there had been a public weighing at the gym. All female applicants had to line up and get weighed, and, if they were over the 115-pound limit, they were rejected without a chance to demonstrate their skills. Janet had been starving herself for days, hoping to make the weight cut, but had failed.

A policy like this sends a clear message—there is an "ideal" body image a woman must conform to if she wants to become a cheerleader. Society expects to find petite women on a college cheerleading squad, "girls" whom male cheerleaders can tumble and lift in cheerleading routines. The cultural message is the same for other popular collegiate groups such as sororities and high status cliques: a thin woman is a "valued" woman.

Compare Janet with her contemporary, Doug Flutie, Boston College's

Heisman trophy-winning football hero. He too fought a cultural belief—that football requires big men. Flutie is only 5' 9" and well below the weight of the typical quarterback. However, through skill and agility, he shattered the stereotype of the tall, rangy quarterback. Cheerleading does not offer a similar opportunity to women. It was possible for Doug Flutie to become a Little Big Man, but not for Janet to become a Big Little Woman.

It is no wonder that American women are obsessed with thinness. They are exhorted to strive for a physical ideal that is laden with moral judgment. Slenderness represents restraint, moderation, and self-control—the virtues of our Puritan heritage. But our culture considers obesity "bad" and ugly. Fat represents moral failure, the inability to delay gratification, poor impulse control, greed, and self-indulgence.

The slim figure has also come to represent health as well as beauty. It is promoted in advertisements for the multimillion-dollar beauty industry, the pharmaceutical industry, and the food industry. Bookstores are full of advice on losing weight, flattening the stomach, getting rid of cellulite, or dressing to look more slender. Ten years ago, there were 300 diet books in print;[3] today, there are countless more. Some of the bestsellers include Dr. Robert Haas' *Eat to Win* (over two million copies in print), Harvey and Marilyn Diamond's *Fit for Life* (three million in print), and William Dufty's *Sugar Blues* (over one million in print).

Since the 1960s, the ideal body type for women has become steadily slimmer and less curvaceous than in the 1950s, which had idolized Marilyn Monroe's bosomy beauty. In the 1970s, visitors to Madame Tussaud's London wax museum rated Twiggy as the "most beautiful woman in the world." [4] Playboy centerfolds and Miss America contestants have become more and more slender between 1958 and 1988, and the actual Miss America winners are thinnest of all.[5]

Fueling this trend are large-scale market interests that exploit women's insecurities about their looks. American food, weight loss, and cosmetic industries thrive on the purchases made to obtain the unobtainable goal of physical perfection. The slim and flawless cover girl is an icon created by capitalism for the sake of profit. Millions of women pay it homage.

But why are women especially vulnerable to eating disorders? Influenced by patriarchal institutions, from the conventional family to schools and the media, girls as young as seven and eight learn that the rewards of our society go to those who conform, not simply on the level of overt behavior, but on the level of biology. If you want to be valued, as a potential spouse, as a coworker, as a friend, then get thin.[6]

To understand this disturbing social phenomenon, I initiated a survey of 395 students (282 females and 113 males) concerning their eating habits, diets, and their attitudes toward self, family, friends, and school. I included interviews in the health care and fitness industries.

I also conducted an in-depth study of 60 college-age women over an eight-year period. I spent an average of three hours interviewing each one, and was able to talk with many of them again during the course of their college careers. I followed a few of these women through graduate school and beyond. (Names and events have been altered to protect each person's anonymity.) Their stories constitute a critical part of this book. Throughout, I chose to use the dramatic metaphor *The Cult of Thinness* because the basic behavior associated with culthood—ritualistic performance and obsession with a goal or ideal—is also characteristic of many modern women.[7] I hope to convey the intense, day-to-day involvement that the pursuit of thinness demands. The body rituals women practice, and the extent to which they sacrifice their bodies and minds to this goal, seem to create a separate reality for its followers. An extensive interview with Anna, a woman who had actually been a member of a religious cult, helped me to see the many parallels (Chapter 1).

In the course of investigating this cult we'll examine why women place such a high premium on their bodies (Chapter 2). We'll see how body perceptions differ among men and women, and how weight has become a primary definer of women's worth and identity. Historically, women have always gone to great lengths to transform themselves to meet the changing cultural requirements of femininity. I trace the mind/body dualism in Western cultural thought, which casts women in the role of the body, and men in the role of in the mind. I show how dominant social and economic interests, sometimes characterized as "patriarchal" and "capitalist," shape this dualism. The essence of the term "patriarchy" is literally "rule of the father."[8] In the context of modern times and for the purposes of this study, patriarchy can be defined as a "system of interrelated social structures, and practices in which men dominate, oppress and exploit women."[9] In this study I will examine the various manifestations of patriarchy as they have evolved within and between social and economic institutions. We will find that one major manifestation of patriarchy is the primary image of women as good wives and mothers and objects of decorative worth. I define the basic nature of capitalism as a political/economic system based on the principle of a competitive, free market economy. These interests have made big business out of women's preoccupation with their bodies. Aided by advertising and mass media, the Cult of Thinness generates enormous profits for the food, diet, and health industries (Chapters 3 and 4).

For me, the most personally upsetting research data comes from the case studies of real women. Judged either by outward appearances, or by the standards set by the Metropolitan Life Insurance's ideal weight chart, few of the students I interviewed would be characterized as overweight. Anyone might look at them and see "normal" successful female co-eds—indeed, the women themselves also commented on this fact, except for those who were clearly anorexic. Yet all were very concerned about their weight, and some were truly obsessed by it. *Their* testimonies show how the Cult of Thinness promotes eating disorders. In Chapters 5, 6, and 7, their voices tell us about their dissatisfactions with their bodies, their concerns with food, and the cultural pressures they feel. They tell us why the pursuit of thinness has become their consuming issue.

For the most part, the women I interviewed were young adults from white upper-middle and middle-class families. But I also discovered how the Cult's message is spreading to other populations: to pre-teens, to men, and to ethnic and social classes that were formerly untouched (Chapter 8).

Ultimately, we need to look for ways to break free from the Cult. It is important to say at the outset that I am not advocating that women give up the pursuit of beauty and fashion entirely. Adornment of the body, and beauty and fashion rituals, form an integral part of all human cultures. But current beauty and fashion trends that advocate attaining ultra-thinness at any price are unrealistic. Given that most women's bodies do not naturally fit the thin ideal, these trends are destructive to women's health, self-esteem, and economic and social advancement within society.

The final chapter of this book evaluates the range of therapies and personal and collective action available to help women overcome their weight obsessions and eating problems. Many of them provide a new framework for envisioning femininity and personal power, for overcoming body insecurity, and for strengthening the inner self. Others involve changing the cultural environment itself.

Throughout my analysis I show examples of women who resist becoming cult victims. Women can abandon their cultural baggage. They can reject the culturally dictated ideal, and transform it to their own image. There *are* alternatives to the Cult of Thinness.

1 A Cult Grows in America

"Ever since I was ten years old, I was just a very vain person. I always wanted to be the thinnest, the prettiest. 'Cause I thought, if I look like this, then I'm going to have so many boyfriends, and guys are going to be so in love with me, and I'll be taken care of for the rest of my life. I'll never have to work, you know?"—Delia, college senior

What's Wrong with This Picture?

Pretty, vivacious, and petite, Delia was a picture of fashionable perfection when she first walked into my office. Her tight blue jeans and fringed Western shirt showed off her thin, 5-ft frame; her black cowboy boots and silver earrings completed a presentation that said, "Look at me!"

The perfect picture had a serious price. Delia had come to me to talk about her "problem." She is bulimic. In secret, she regularly binges on large amounts of food, then forces herself to vomit. It has become a powerful habit, one that she is afraid to break because it so efficiently maintains her thin body. For Delia, as for so many others, being thin is everything.

"I mean, how many bumper stickers have you seen that say 'No Fat Chicks,' you know? Guys don't like fat girls. Guys like little girls. I guess because it makes them feel bigger and, you know, they want somebody who looks pretty. Pretty to me is you have to be thin and you have to have like good facial features. It's both. My final affirmation of myself is how many guys look at me when I go into a bar. How many guys pick up on me. What my boyfriend thinks about me."

Delia's Story

Delia is the eldest child, and only girl, in a wealthy Southern family. Her father is a successful dentist and her mother has never worked outside the home. They fought a lot when she was young—her father was an alcoholic— and they eventually divorced. According to Delia, both parents doted on her.

"I've never been deprived of anything in my entire life. I was spoiled, I guess, because I've never felt any pressure from my parents to do anything. My Dad would say, 'Whatever you want to do, if you want to go to Europe, if you want to go to law school, if you don't want to do anything . . . whatever you want to do, just be happy.' No pressure."

He was unconcerned about her weight, she said, but emphasized how important it was to be pretty. Delia quickly noticed this message everywhere, especially in the media.

"I am so affected by *Glamour* magazine and *Vogue* and all that, because that's a line of work I want to get into. I'm looking at all these beautiful women. They're thin. I want to be just as beautiful. I want to be just as thin. Because that is what guys like."

When I asked what her mother wanted for her, she recited, "To be nice and pretty and sweet and thin and popular and smart and successful and have everything that I could ever want and just to be happy." "Sweet and pretty and thin" meant that from the age of ten she was enrolled in a health club, and learned to count calories. Her mom, who at 45 is "beautiful, gorgeous, thin," gave her instructions on how to eat.

" 'Only eat small amounts. Eat a thousand calories a day; don't overeat.' My mom was never critical like, 'You're fat.' But one time, I went on a camping trip and I gained four pounds and she said, 'You've got to lose weight.' I mean, she watched what I ate. Like if I was going to get a piece of cake she would be, 'Don't eat that.' "

At age 13 she started her secret bingeing and vomiting. "When I first threw up I thought, well, it's so easy," she told me. "I can eat and not get the calories and not gain weight. And I was modeling at the time, and I wanted to look like the girls in the magazines."

Delia's preoccupation with thinness intensified when she entered high school. She wanted to be a cheerleader, and she was tiny enough to make it. "When I was sixteen I just got into this image thing, like tiny, thin . . . I started working out more. I was Joe Healthy Thin Exercise Queen and I'd just fight eating because I was working out all the time, you know? And so I'm going to aerobics two or three times a day sometimes, eating only salad and a bagel, and like, no fat. I just got caught up in this circle."

College in New England brought a new set of social pressures. She couldn't go running every day because of the cold. She hated the school gym, stopped working out, and gained four pounds her freshman year. Her greatest stress at college had nothing to do with academics. "The most stressful thing for me is whether I'm going to eat that day, and what am I going to eat," she told me, "more than getting good grades."

After freshman year Delia became a cheerleader again. "Going in, I know I weighed like 93 or 94 pounds, which to me was this enormous hang-up, because I'd never weighed more than 90 pounds in my entire life. And I was really freaked out. I knew people were going to be looking at me in the crowd and I'm like, I've got to lose this weight. So I would just not eat, work out all the time. I loved being on the squad, but my partner was a real jerk. He would never work out, and when we would do lifts he'd always be, 'Delia, go run. Go run, you're too heavy.' I hadn't been eating that day. I had already run seven or eight miles and he told me to run again. And I was surrounded by girls who were all so concerned about their weight, and it was just really this horrible situation."

College life also confirmed another issue for Delia, a cultural message from her earliest childhood. She did *not* want to be a breadwinner. She put it this way, "When I was eight I wanted to be President of the United States. As I grew older and got to college I was like, wow, it's hard for women. I mean, I don't care what people say. If they say the society's liberated, they're wrong. It's still really hard for women. It's like they look through a glass window [*sic*]. They're vice presidents, but they aren't the president. And I just figured, God, how much easier would it be for me to get married to somebody I know is going to make a lot of money and just be taken care of . . . I want somebody else to be the millionaire."

Delia said she lived by three simple rules. To the Duchess of Windsor's dictum, "You can never be too thin or too rich," she added "Be confident and be funny," and "You eat to live not live to eat." She ignored the fact that *not* eating, or getting rid of what she *had* eaten, controlled her life.

The Cult of Thinness

Delia's single-minded pursuit of thinness and beauty has many parallels to a religious cult. In both cases a group of individuals is committed to a life defined by a rigid set of values and rules. Members of true cults frequently isolate themselves from the rest of the world and develop a strong sense of community. They seem obsessed with the path to perfection, which, though unattainable, holds out compelling promises. In following their ideals, they usually feel that they are among "the chosen."

To get an idea of what members of true cults experience, I interviewed Anna, a psychology student in her forties who had been a member of a spiritual cult many years earlier. Her descriptions of the required total immersion in a separate and controlled reality were chilling.

"We really became this very separate group of people, because our leader made that happen. He separated us from our families, and made us feel very bound to each other and to him, and different from everybody else."

But remaining in those ranks means constant vigilance. It demands adherence to certain beliefs (which may seem quite bizarre to outsiders) and the practice of specific rituals. Most religious cults center on a spiritual leader who defines the path, and who threatens exile or worse to those of his flock who stray.

"Our guru considered himself infallible," Anna said. "He was beyond feedback, so that if I looked at him and saw something I didn't like, he told me that I was projecting, that he was merely a mirror for my own shortcomings . . . He was really good at playing upon your weaknesses as a way to keep people bound to him, and I felt that the only reason my life had value was because of him and what this way of life had given me. Subconsciously, I thought I would die if I left him."

Young women like Delia invest in thinness with the same intense, moment-to-moment, day-to-day involvement as religious cult members. They may not answer to a single leader, but bow instead to powerful cultural forces that define females in terms of their physical attributes. The influence of these forces is so pervasive that, in many ways, it is harder to resist.

Being female is the primary criterion for membership in the Cult of Thinness. The object of worship is the "perfect" body. The primary rituals are dieting and exercising with obsessive attention to monitoring progress—weighing the body at least once a day and constantly checking calories. The advertising industry and the media provide plenty of beautiful-body icons to worship. There are numerous ceremonies— pageants and contests—that affirm the ideal.

And there are plenty of guides and gurus along the way. Often it is the mother who initiates the young novice into the secrets of losing weight. Some of the most revered oracles have celebrity as their major qualification. Jane Fonda, Cindy Crawford, and Oprah Winfrey are among those who advise their fans on the virtues or pitfalls of certain diets and exercises.

Other sages have medical qualifications and have produced "sacred texts." Dr. Atkin's diet, the Scarsdale Diet, and the Pritikin Diet were created by best-selling diet doctors. The diet gurus can also be psychologists. Some make motivational audiotapes for their patients. Others have special phone-in hours for those who have fallen off their diets, or provide a special "intensive care" line for those in dire need.[1] Such gurus come and go.

Diet clubs and twelve-step weight loss programs introduce even more fervor toward shedding pounds. Their meetings are filled with conversion stories of how so-and-so "saw the light" and lost pounds, or fell from grace by eating "forbidden" foods. There are recommended penances, as well as weekly support groups and telephone chains to help the backsliders. One researcher who attended over 90 group diet meetings compares them to quasi-religious experiences. She notes:

> These dieters labeled overweight as a sinful deviation, buttressed by the religious argot of saint, sinner, angel, devil, guilt, confession and absolution. Some stated

that they had innocently caught this sin, as rather passive and helpless illness victims. Others claimed that they had actively acquired their sinful state—as "almost criminals," hanging their heads in shame, they assumed the responsibility for their fat "misdemeanors" or "felonies."[2]

Those who have experienced the shared sacrifice of the cult create a "sacred" environment. Their common lifestyle brings them together and drives a wedge between them and the rest of society, who may come to be viewed as "profane." This split between what is considered sacred and profane mimics what our society associates with the terms "thin" and "fat."

Thin is sacred. Thin is beautiful and healthy; thin will make you happy. If you are female, thin will get you a husband. Salvation awaits those who attain the ideal body.

Fat is profane. To be fat is to be ugly, weak, and slovenly; to have lost control, be lazy, and have no ambition. Achieving the proper weight is not just a personal responsibility, it is a moral obligation. Those who indulge in gluttony and sloth do not want to be among the "saved."[3]

Just as there is a range of faith among the devotees of any religion, the women I interviewed can be positioned along a continuum. The most avid members of the Cult of Thinness engage in practices more characteristic of fringe cult movements, like the followers of Reverend Moon or Jim Jones. The rituals surrounding anorexia, bulimia, and exercise addiction carry the risk of emotional and physical damage, or even death.

A Dangerous Tool

Members of the Cult of Thinness use whatever strategies they can find to strive for the ideal body. Delia believed that it was her appearance that would lead to her goal of marriage to a millionaire. Her eating disorder was a tool for accomplishing this end.[4]

The concept of "tool" is important to our understanding of the increase in eating disorders among American females. Delia's behavior was a straightforward means of meeting the stringent standards of beauty dictated by her culture. What a clinician would view as pathological, Delia saw as utilitarian, along the same lines as cosmetic surgery. When I asked her how she was going to deal with her eating "problem" she replied, "I have no idea. I mean, I don't see it going away. I guess it's kind of sick. Once I lose five pounds, I'll be really happy with my body, but I will always be working and eating right and exercising to keep the five pounds off because that is very, very important to me. And however contradictory and sick and overrated and vain and anything else you can say about it, it's the truth. My bulimia, my exercising and even abstaining from food, they're tools for me to lose weight."

An advertisement in a recent issue of *Teen* magazine illustrates the different ways women can "remodel" their looks with specific tools or "building" supplies from Rubbermaid®. A woman, who is shown from the nose up, is curling her hair with a curling brush from Rubbermaid®. Our attention is drawn directly to her eyes, which display an intense focus on curling her hair and, therefore, improving her looks. The accompanying caption reads, "If this is your remodeling job, here's your tool kit." Beauty is depicted not only as a job, but as a lifetime career that requires time, energy, and, above all, the proper tools, which Rubbermaid® will willingly supply. Women are encouraged to focus on their bodies; there is no mention of furthering the intellect or careers. Compare that image to an ad from *MacUser* magazine (Figure 1) whose readership is not predominantly female. A computer device displayed on a construction worker's tool belt sends a message about achievement in the male world of work—the world of the mind, not the body.

Economic and career achievement is a primary definition of success for men. (Of course, men can also exhibit some self-destructive behaviors in pursuit of this success, such as workaholism or substance abuse.) Delia's upbringing and environment defined success for her in a different way. She was not interested in having a job that earned $150,000 a year, but in marrying the guy who did. She learned to use any tool she could to stay thin, to look good, and to have a shot at her goal.

No wonder she was reluctant to give up her behavior. She was terrified of losing the important benefits of her membership in the Cult of Thinness. She knew she was hurting psychologically and physically, but, in the final analysis, being counted among "the chosen" justified the pain.

"God forbid anybody else gets stuck in this trap. But I'm already there, and I don't really see myself getting out, because I'm just so obsessed with how I look. I get personal satisfaction from looking thin, and receiving attention from guys."

I told Delia about women who have suggested other ways of coping with weight issues. There are even those who advocate fat liberation, or who suggest that fat is beautiful. She was emphatic about these solutions.

"Bullshit. They live in la-la land . . . I can hold onto my boyfriend because he doesn't need to look anywhere else. The bottom line is that appearance counts. And you can sit here and go, 'I feel good about myself twenty pounds heavier,' but who is the guy going to date?"

A Woman's Sense of Worth

Delia's devotion to the rituals of beauty work involved a great deal of time and energy. She weighed herself three times a day. She paid attention to what

Network nitty-gritty.

Figure 1 Farallon, PhoneNET, September, 1989

she put in her mouth; when she had too much, she knew she must get rid of it. She had to act and look a certain way, buy the right clothes, the right makeup. She also watched out for other women who might jeopardize her chances as they vied for the rewards of the system.

A woman's sense of worth in our culture is still greatly determined by her ability to attract a man. Social status is largely a function of income and occupation. Women's access to these resources is generally indirect, through marriage.[5] Even a woman with a successful and lucrative career may fear that her success comes at the expense of her femininity.

The way Hillary Rodham Clinton's image evolved during the 1993 presidential election campaign is a good example. A graduate of Yale Law School, she was a partner with a prestigious law firm and made more money than her husband. Yet on the campaign trail, many were reportedly "put off by her assertiveness." To counter this negative impression, the campaign strategists softened her image to portray her as the "woman behind the man." Hillary Clinton began to appear more often with her child, and photos frequently showed her alongside or behind her husband.[6] She followed advice on her clothes, her hair, and her overall appearance. Even women who have direct access to high social status do not escape more mundane interpretations of attractiveness and femininity.

Body weight plays an important part in physical attraction. In research studies that asked people what attributes are most indicative of "positive appearance," weight was a key factor.[7]

As women increasingly enter the labor market, weight again may have a role in determining their desirability for certain jobs.[8] For example, until recently, the guidelines for flight attendants (a job considered to be women's work) required a 5-foot-5 inch female flight attendant to weigh 129 pounds or less. Every year dozens of flight attendants who could not meet these weight guidelines were fired. This begs the question: Who makes the rules? Is safety really the only guideline? Can't a flight attendant be agile at 140 or 150 pounds? Is a weight limit a legitimate way of ensuring a safe flight? The Equal Employment Opportunity Commission (EEOC) and the Association of Professional Flight Attendants challenged the guidelines with a lawsuit, but a weight limit, though higher, continues to be in effect.

Cultural messages on the rewards of thinness and the punishments of obesity are everywhere. Most women accept society's standards of beauty as "the way things are," even though these standards may undermine self-image, self-esteem, or physical well-being. Weight concerns or even obsessions are so common among women and girls that they escape notice. Dieting is not considered abnormal behavior, even among women who are not overweight. But only a thin line separates "normal" dieting from an eating disorder.[9]

Theories of Eating Disorders

There are several ways to view eating disorders. Early theories rely on what I have termed "individualistic" explanations.

One theory assumes that an eating disorder is a reflection of a woman's psycho-sexual development. Women with eating disorders are said to fear oral impregnation and reject their sexuality.[10] Refusing food is one of the ways that an adolescent girl can gain some control and autonomy over the frightening changes in her body and her life. Psychotherapy is usually the treatment.

Another view has focused on biological causes. This view links eating disorders with depression, which may be caused by a chemical or metabolic defect, especially in women.[11] Drugs, sometimes in the form of hormones, are considered a dependable, quick, and inexpensive treatment.

Relatively recent thinking sees an eating disorder as the result of family dynamics. A power struggle between child and parent, especially the mother, leads to rejection of the mother as a role model.[12] This theory emphasizes relations between people rather than conflict within a person, and places little importance on wider factors outside a family unit. It is treated with family therapy.

These alternative views imply that the solution to an eating disorder lies within the individual or the family unit. Since the problem centers there, it is also the target for change. Clearly, it is important to help individuals or families overcome their personality and even chemical "deficits" by identifying those at risk. But this approach often amounts to "blaming the victim."[13]

In Delia's history, evidence of psychological trauma might help us understand how her eating problem arose. She mentioned her dysfunctional family life when she was growing up: "I think I first threw up in the hope of getting attention, because that was the year before my dad went into treatment for his alcoholism. Things were really bad at my house, as far as my parents fighting. If I could catch their attention that I was throwing up, maybe they wouldn't fight. I didn't feel fat at all."

But it didn't have that effect. Amid the emotional turmoil at home, she discovered the potent physical comfort of bingeing. By making herself vomit, she found a way to "eat her cake and get rid of it too." And she learned to keep it a secret. "Nobody knew."

Why Here, Why Now?

There *are* psychological reasons for some of the eating problems occurring in young college women like Delia. Yet that perspective does not answer deeper questions: Why does Delia express her psychological distress through her body? If she'd grown up in another decade, would her distress manifest itself in another way, like Victorian hysteria, for example? Why does she continue to pursue this behavior, even when Mom and Dad have resolved their marital issues and her father has been a recovering alcoholic for many years? Why has this problem mushroomed at this point in history for so many women of her race, age, and class?

We need to look beyond individual factors and examine the wider environment. The current outbreak of eating problems especially among white, middle-class, college-age women is a reflection of a Cult of Thinness. Social and economic forces pressure Delia and others to pursue the thinness ideal, even to the point of dangerous behavior. What makes this pursuit so rewarding? How can it seem so "normal"? Who benefits from this Cult of Thinness? Looking at those influences, rather than individual factors, will help us understand the origins of this cult and why it has spread so rapidly. Perhaps it will suggest a new solution to the problem of women's relations to their bodies and to food.

2 Men and Women: Mind and Body

"You know guys just sort of think you're stupid. Sure, I get a lot of attention being blonde and female, and my idea of what society really thinks of women is that intelligence doesn't even figure in. Just body and face. I remember the time my mom asked me what I was going to be when I grew up. I said if I could be anything—anything in the world—I'd put myself on the cover of a magazine. For men, it's the money they make, not how they look. Nobody cares if a guy is a good father or even a good person. People say, Oh, yeah, he's the president of this company or that, or the head doctor of such and such a hospital. It's his job that counts."—Tracy, college sophomore.

Our culture judges a man primarily in terms of how powerful, ambitious, aggressive, and dominant he is in the worlds of thought and action. These are qualities more of the mind than the body. A woman, on the other hand, is judged almost entirely in terms of her appearance, her attractiveness to men, and her ability to keep the species going. Her sexual body can even be quite dangerous to a man on his way to success. She becomes a temptress distracting him from his true work, the pursuit of rationality, knowledge, and power.

The split between mind and body is a central idea in Western culture. It often frames our perceptions of what it means to be feminine and masculine.[1] This has been illustrated by studies conducted by Paul Rosenkrantz and his colleagues in the 1960s.[2] They gave college students a list of extreme personality traits and their opposites: very passive vs. very active; very illogical vs. very logical; very vain about appearance vs. uninterested in appearance, and so on. The students were asked to ascribe each trait to either males or females, and to rank each trait's social desirability. The findings from this study (and from others) showed that traits most often associated with competence and social desirability were assigned to men and those associated with having an "emotional" life were assigned to women. Men were viewed as more independent than women, more logical, more direct, more self-confident, more ambitious. Women were seen as being more gentle than men, more softly spoken, more talkative, and more tender.

The problem with a dichotomy is that it provides no middle ground. If one trait is positive, then its opposite is negative—if a man is strong, a woman is weak. But such a narrow view of human behavior ignores the reality that there are a range of traits common to both sexes.[3]

The college-age women I interviewed fully understood this mind and body gender stereotyping. Angela said: "My body is the most important thing. In a way, it's like that's all I ever had. Just because that's all everyone ever said about me. My mother would say that I am smart and stuff, but really they focused on my looks. And even my doctor enjoys my looks. He used to make me walk across the room to check my spine and he'd comment on how cute I walked, that I wiggled. Why comment on it at all?"

She noted how difficult it is for women to be acknowledged for "male" attributes.

"I think men are shocked when they see a gorgeous, really intelligent woman. You'll always hear about how 'gorgeous.' You don't hear or very rarely hear, 'She's a very intelligent woman.' "

Historical Roots of the Mind/Body Dichotomy

The notion of a split between mind and body dates back at least to ancient Greece. Aristotle, in the fourth century B.C., asserted that males were superior to women, whom he described as "monsters . . . deviated from the generic human type." Women were imperfect versions of the ideal form of humankind, "mutilated males," who were emotional and passive prisoners of their body functions.[4]

Scholar Donna Wilshire explains that Aristotle's world contained a variety of hierarchical dualisms: soul ruled the body, reason was preferable to emotion. The Mind, which only a male could have, was said to be connected to the "divine" soul. The female, therefore, was incapable of reason:

"For him, Pure Mind ['Nous,' only possible for males] is connected with 'divine' Soul, which is supreme of all earthly things. The male Mind is therefore *higher* and *holier* than all matter, even higher than the beloved Apollonian [ideal, male] body; certainly the male Mind and Reason rule over and are 'more divine' than the female body because she [being ruled by emotions and body functions] is not capable of Mind or Reason, and so on."[5]

Even Aristotle's theory of embryology states that all life (soul) is contained in the man's semen. He thought that the mother was merely a receptacle. All "true" babies were male babies and the birth of a female child was considered a failure in the birth process.[6]

Aristotle's work powerfully influenced Christian doctrine. St. Augustine in the fifth century and St. Thomas Aquinas in the thirteenth century asserted the primacy of male over female. Thomas Aquinas stated, "As regards the individual nature, woman is defective and misbegotten, for the active force in the male seed tends to the production of a perfect likeness in the masculine sex."[7] His teacher, Albert the Great, went even further: "One must be on one's guard with every woman, as if she was a poisonous snake and the horned devil. . . . Her feelings drive woman towards every evil, just as reason implies man toward all good."[8] Her redemption from such evil did not lie in the development of her capacity to think, but in her duties of wife and mother.[9]

During the seventeenth century, the period of the Enlightenment, knowledge required "objectivity."[10] A scientist should be detached from all emotional and personal considerations in order to ascertain "the truth." The very notion of the dispassionate scientist, whose mind transcended his body, defined science as a male pursuit.[11] The object of scientific knowledge—that is, nature—was female.

Sir Francis Bacon's description of the scientific process uses imagery that suggests a man dominating a woman:

> You have but to hound nature in her wanderings and you will be able when you like to lead and drive her afterwards to the same place again. Neither ought a man to make scruple of entering and penetrating into those holes and corners when the inquisition of truth is his whole object.[12]

Because the body was considered to be of a lower order and in fact could interfere with the pursuit of the truth, women were viewed as incapable of scientific thinking. The medical literature of the time supported this view. It is replete with examples of the mind/body dichotomy, especially in theories of reproduction. During the nineteenth century, scientists thought any type of mental work was too taxing to the female reproductive organs: "Victorian scientists promoted the 'cult of true womanhood,' discouraged education for women because too much of women's energy would go to their brains, causing their reproductive organs to atrophy, and asserted that menstruation results in irrationality and loss of mental powers."[13]

Cultural rules have controlled women's bodies throughout history.[14] Our Anglo-American legal institutions, for example, created laws based on the biological differences between men and women. United States laws enacted at the turn of the century regulated the number of hours per day or week that women were allowed to work. Other laws prevented them from working at night or in certain occupations like mining or smelting. This "protective legislation" drove women out of certain jobs, while it gained reduced hours for men in those occupations where women remained. In addition, domestic-relations laws (those concerning property rights, pension benefits, maternity leave policies, etc.) have reinforced the idea that women are reproducers (the reproductive body) and men are breadwinners (the rational mind).[15]

For those of us within a particular society, it is sometimes difficult to see how culture controls women's bodies. In America today, women who diet, or have their breasts enlarged and their tummies tucked, regard this as an exercise of free will. But if we compare these practices with two historical examples, one from ancient China and the other from the Victorian era, we may gain a new perspective.

Ancient China and the Practice of Foot Binding

For one thousand years, the Confucian philosophy of patriarchal authority formed the basis of Chinese society. It was a hierarchical system. Males were considered higher than females, and the old had authority over the young. As Confucius, born in 551 B.C., wrote, "Women are, indeed, human beings, but they are of a lower state than men."[16]

Foot binding, one of the most dramatic examples of control over such "lower beings," developed in pre-revolutionary China. The custom originated around the tenth century, with court dancers who wrapped their feet to imitate pointed, sickle moons (not unlike toe shoes in Western ballet). The Chinese court and the upper class had always prized small feet in women: now they copied this practice and took it to extremes. It became an important symbol of high status within Chinese society. In time, it filtered down to the masses as well.[17]

This custom lasted more than a thousand years and served to virtually cripple women in the name of beauty and femininity. Little girls had their toes bent under into the sole, with the heel and toes bowed forcibly together, and wrapped tightly. The bones eventually broke and the foot could no longer grow. This severely deformed clubfoot, only a few inches long, became known as the "lotus" or "lily" foot. Walking on these stumps was painful, if not impossible.

One young girl who had to endure this process and suffered intensely wrote:

> Born into an old-fashioned family at P'ing-hsi, I was inflicted with the pain of foot binding when I was seven years old. I was an active child who liked to jump about, but from then on my free and optimistic nature vanished. Binding started in the second lunar month; mother consulted references in order to select an auspicious day for it. I wept and hid in a neighbor's home, but mother found me, scolded me, and dragged me home. She shut the bedroom door, boiled water, and from a box withdrew binding, shoes, knife, needle and thread. I begged for a one-

day postponement, but mother refused. "Today is a lucky day," she said. "If bound today, your feet will never hurt; if bound tomorrow, they will." She washed and placed alum on my feet and cut the toenails. She then bent my toes toward the plantar [the mid-region of the sole] with a binding cloth ten feet long and two inches wide, doing the right foot first, and then the left. She finished binding and ordered me to walk, but when I did the pain proved unbearable.

That night, mother wouldn't let me remove the shoes. My feet felt on fire and I couldn't sleep, mother struck me for crying. On the following days, I tried to hide but was forced to walk on my feet. Mother hit me on my hands and feet for resisting. Beatings and curses were my lot for covertly loosening the wrappings . . . Mother would remove the bindings and wipe the blood and pus which dripped from my feet. She told me that only with removal of the flesh could my feet become slender.

Every two weeks, I changed to new shoes. Each new pair was one- to two-tenths of an inch smaller than the previous one. The shoes were unyielding and it took pressure to get into them . . . After changing more than ten pairs of shoes, my feet were reduced to a little over four inches. I had been binding for a month when my younger sister started; when no one was around we would weep together.[18]

Why did women, especially mothers, continue a custom that inflicted such suffering?[19]

There were several reasons. The bound foot, a symbol of feminine beauty, represented a woman's only prospects in life. Inheritance laws of Chinese society were male dominated. Unable to inherit property or pass on the ancestral name, a girl was an economic liability until she left to join her husband's family. A good match with prosperous in-laws offered a girl's parents a chance to recoup their investment in raising a daughter.[20] Her only value lay in her marriageability. While the bride's parents could gain social and economic status, the groom's parents acquired another source of labor, both productive and reproductive. The type of work she provided varied by social class. Upper-class women did not work outside their homes, but even though footbound, they still performed household chores, made handicrafts, and attended to the needs of their extended families.[21] Even peasant families who dreamed of marrying into a higher class crippled their daughters accordingly. Only the poorest female field workers escaped. They could not afford to have their feet more than moderately bound.[22]

This custom lasted as long as it did because it reinforced the patriarchal authority of Chinese society. It supported dualistic thinking (men were superior and women inferior; men were valued and women devalued). Women followed a strict line of obedience, first to fathers, then to husbands, and finally to sons upon the death of the husband.[23] Foot binding was part of society's belief in the dichotomy between mind and body. Control over women could be justified in this way:

The minds of footbound women were as contracted as their feet. Daughters were taught to cook, supervise the household, and embroider shoes for the golden Lotus. Intellectual and physical restriction had the usual male justification. Women were perverse and sinful, lewd and lascivious, if left to develop naturally. The Chinese believed that being born a woman was payment for evils committed in a previous life. Footbinding was designed to spare a woman the disaster of another such incarnation.[24]

Foot binding also restricted female mobility and sexuality, which are potential sources of power and resistance for women in pre-industrial societies. As Susan Greenhalgh notes in her research on women in Old China,

> The greatest threat to the family system came from the women because they married in from an *outside* family. Women marrying into the patriarchal family could disrupt its stability by offering dissenting opinions about the allocation of labor and goods within the family, or by simply refusing to accept patterns of authority and interaction already established, and returning to their natal homes.[25]

Girls had to embrace a set of personality traits along with their bound feet. These traits defined womanly behavior as "chaste and yielding, calm and upright." A woman's speech should be "not talkative yet agreeable"; her appearance and demeanor both "restrained and exquisite"; and she should engage in work that demonstrated her skills in "handiwork and embroidery."[26] Upper-class Chinese women, who spent a great deal of time and energy (and pain) on this feminine ideal, were unlikely to challenge the established order.

Sexuality, another threat, also had to be channeled. Binding limited the possibility of extramarital affairs, since the bound woman had to be carried over any distance and could not leave the home unaided. One governor of the Chang Prefecture, in Fukien, felt that "women there tended to be unchaste and to indulge in lewdness."[27] He ordered that all women in the town must have their feet bound excessively. They could get about only by leaning on canes. "Whenever they attended local celebrations or funerals, such gatherings were called 'A Forest of Canes.' Their bound feet were smaller than the norm; this was attributed to the desire to prevent them from eloping."[28]

Many even believed that foot binding itself helped create an ideal female figure. One Taiwanese doctor actually wrote:

> Foot binding had a physical influence on a woman's body. When the foot-bound woman went walking, the lower part of her body was in a state of tension. This caused the skin and flesh of her legs and also the skin and flesh of her vagina to become tighter. The woman's buttocks, as a result of walking, became larger and more attractive sexually to the male.[29]

And finally, as a prominent aspect of feminine identity, the bound foot was an important part of sexual rituals—even a focal point of sexual excitement. "The tiny and fragile appearance of the foot aroused in the male a combination of lust and pity. He longed to touch it, and being allowed to do so meant that the woman was his."[30]

The practice of foot binding in ancient China seems grotesque and cruel. But it can help us understand the current Cult of Thinness. Foot binding reflected the economic and social power structure within a certain patriarchal society. Societies based on such authority hold up a mirror that defined women in terms of their bodies. They are "commodities" for domestic economic exchange and social control.

When societies make the transition from medieval or traditional political-economic systems, like ancient China, to modern systems, power within that society shifts. It moves from centralized authority to a diffusion of power among various institutions, including those controlled by patriarchal interests.[31] Capitalism is a dominant force within modern society. Early capitalism relied on the *external* control of women's bodies. One example, as we will see, was the practice of corseting.[32] But over time, this control has become more *internal*, through self-imposed body practices and rituals. Because modern women are also consumers, this focus on their bodies has created many multibillion dollar industries.

The Rise of Consumer Culture and the Practice of Corseting

Women wore the corset in England and America for most of the nineteenth century. Corseting demonstrates women's emerging roles as both consumer and commodity during the rise of capitalism.

The image of the Victorian woman was, in part, a response to the dramatic changes that accompanied industrialization. In this time of vast economic expansion, enough people became wealthy to form a large, prosperous middle class. Work became segregated from the home, where the white middle-class woman was expected to stay, supported by a well-to-do husband.

She was also expected to be a fragile, pure creature, submissive to her spouse and subservient to domestic needs. Her decorative value, a quality that embraced her beauty, her character, and her temperament, defined her worth. This ideal, later referred to as the "cult of true womanhood," demanded uncompromising virtue.[33] Like the rich woman of ancient China, the Victorian woman became a prized showpiece of her husband's wealth. She was the manager of hearth and home. Middle- and upper-class wives became the chief consumers of early capitalists' products.[34]

Instead of the clubbed lily foot, the important symbols of beauty and status for women were a tightly cinched waist, paleness, and fragility. The waist had a special erotic significance; it symbolized passivity, dependence, and, more perversely, bondage.[35] The proud husband encircling his wife's waist in his broad hands, notes one researcher, demonstrated power and control.[36] A French beauty writer, Mlle. Pauline Mariette, had this to say about the vital importance of the waist in defining a woman's body image:

> The waist is the most essential and principal part of the woman's body, with respect to the figure . . . The bee, the wasp . . . those are the beings whose graceful and slender waist is always given as the point of comparison . . . The waist gives woman her jauntiness, the pride of her appearance, the delicacy and grandeur of her gait, the unconstraint and delight of her pose."[37]

Small, slender hands and narrow feet complemented the tiny waist.[38]

To attain such an ideal, a tightly laced undergarment reinforced with whalebone, and later steel, constricted women's waists for many hours a day.[39] This pressure often caused pain and distorted the internal organs and rib cage. One woman described the following ritual:

> I went and ordered a pair of stays, made very strong and filled with stiff bone, measuring only fourteen inches round the waist. These, with the assistance of my maid, I put on, and managed to lace my waist to eighteen inches. At night I slept in my corset without loosing the lace in the least. The next day my maid got my waist to seventeen inches, and so on, an inch every day, until she got them to meet. I wore them regularly without ever taking them off, having them tightened afresh every day, as the laces might stretch a little. They did not open in front, so that I could not undo them if I had wanted. For the first few days, the pain was very great, but as soon as the stays were laced close, and I had worn them so for a few days, I began to care nothing about it, and in a month or so I would not have taken them off on any account. For I quite enjoyed the sensation, and when I let my husband see me in a dress to fit I was amply repaid for my trouble.[40]

A tight corset prevented women from moving around very much, which tended to make them dependent and submissive.[41] Like the family furniture, a wife was considered a possession. Indeed, many were nearly as immobile as furniture. "Accounts of nineteenth-century house fires reveal that women occasionally went up in flames with the household goods because of immobilizing corsets and skirts too full or too tight to run in."[42] In 1867, one English women's magazine reported that 3,000 women were burned alive that year and about 20,000 women suffered severe burns and injuries because they wore crinolines.[43]

Why did women corset themselves so willingly? Like the women of ancient China, their body image largely determined their identity and the rewards they received. As one historian commented, "In an age when alternatives to marriage for women were grim and good husbands scarce, the pressures to conform to the submissive ideal that men demanded were enormous."[44]

One report described what happened when a zealous mother laced her daughter's stays too tight and killed her at the age of 20.[45] "her ribs had grown into her liver, and that her other entrails were much hurt by being crushed together with her stays, which her mother had ordered to be twitched so straight that it often brought tears into her eyes whilst the maid was dressing her."[46]

Patriarchal interests, which characterized women primarily as wives, mothers, and decorative objects, complemented an economy relying more and more on domestic consumption.[47] Capitalism motivated producers to create new needs and exploit new markets, most of which centered around the body and its functioning. Advertising was also crucial in helping to define women as the primary consumers. It promoted insecurity by encouraging women "to adopt a critical attitude toward body, self and life style." They rushed to purchase the latest household items, which were important symbols of being a good wife and mother. They flocked to buy beauty products, which were signs of a woman's femininity and ability to hold on to her man. As long as a woman viewed her body as an object, she was controllable and profitable.[48]

The Origins of the Cult of Thinness: From External to Internal Body Control

Nineteenth-century industrialization and mass production influenced body image in general, for both sexes. In her book, *Never Too Thin*, social historian Roberta Seid points out that, for women, beauty was becoming democratized as ready-made clothing introduced the idea of standard sizes.[49] The machine age promoted a streamlined aesthetic. "While . . . slenderness had been associated with sickness and fragility, now many health authorities cautioned against overeating and excess weight."[50] New studies related obesity to premature mortality. By the turn of the century, technological innovation, economic growth, and efficiency reinforced the ideal of a slender body. These economic imperatives provided the metaphors that shaped standards for what was considered the desired human body: "to be as efficient, as effective, as economical, as beautiful as the sleek new machines, as the rationalized workplace . . . It was these . . . developments that forged the society we know today and that established the framework for our prejudice against fat."[51] In general, Seid notes that men were not bound by the pressure to look slender. A new male image—the "self-made" man—arose with the coming of the Industrial Revolution and the Protestant ethic. The "self-made" man strives for upward mobility through hard work, ability, and thrifty ways, not relying on his physical appearance.[52] In fact men's clothing remained relatively unchanged over many decades.[53] In his book *Fashion Culture and Identity*, sociologist Fred Davis notes, "The restricted character of men's dress code derived principally . . . from the overweening centrality accorded work, career and occupational success for male identity; so much so that for many decades to come, especially in the middle class, clothing was almost unavailable as a visual means for men to express other sides of their personality." [54]

Twentieth-century capitalism includes the diet, beauty, cosmetic, fitness, and health industries. Along with modern patriarchy, it continues to control women through pressures to be thin.[55] The modern woman, however, achieves this new ideal not through the purchase of a corset or girdle, but through self-directed action such as dieting and exercise. Taken to extreme, the logical conclusion of this self-direction is an eating disorder.[56]

The shift from external to internal control was part of the ideology of "women's independence." As the nineteenth century drew to a close, middle class women were increasingly involved in social reform, volunteer activities, and work outside the home such as teaching and nursing.[57] By 1870 more were entering college, and by the 1890s they were beginning to compete with men in such professions as law, medicine,[58] and journalism. A new interest in physical fitness led doctors to prescribe women's tennis, golf, swimming, horseback riding, and bicycling. Dancing was the rage. Suffragettes were marching for the right to vote. Women were becoming more physically mobile and abandoning their tight corsets. [59]

Of course, the rise of the women's movement during the 1920s had a predictable backlash. Women's independence threatens the traditional view of how men and women focus their lives.

> While men are busy conquering and controlling nature and women, women are obsessed with controlling their bodies. Man believes he survives through his enduring achievement. Women is her mortal body . . . A man may sweat, scar and age; none of these indications of physicality and mortality are seen to define the male self. Indeed, those men who take unseemly interest in the body are described as womanly and are presumed to be homosexual.[60]

Just as women were demanding more "space" and more equality, the culture's standards of attractiveness demanded that they shrink.[61] A slender female body, achieved though dieting, has become the dominant image for most of the twentieth century.[62]

In effect, patriarchal and consumer interests co-opted this newfound independence and harnessed it for their own ends. In order to attain the ultra-slender ideal, women began to purchase diet products and to spend enormous amounts of time and energy on their bodies.[63] These activities continue to divert economic and emotional capital away from other investments women might make, like political activism, education, and careers.

These investments could empower women, and change their thinking about mind and body.

Even during the first wave of feminism, the slim, youthful, albeit rather sexless "flapper" of the 1920s became the most important symbol of American beauty. She had a straight, boyish figure, and exposed her slender legs. As one historian notes, it was a trivialized image.

> On the one hand, she indicated a new freedom in sensual expression by shortening her skirts and discarding her corsets. On the other hand, she bound her breasts, ideally had a small face and lips . . . and expressed her sensuality not through eroticism, but through constant, vibrant movement . . . The name "flapper" itself bore overtones of the ridiculous. Drawing from a style of flapping galoshes popular among young women before the war, it connoted irrelevant movement and raised the specter of seals with black flapping paws.[64]

Psychiatrist John A. Ryle observed that it was during the Flapper Era that cases of anorexia nervosa increased, which he attributed to "the spreading of the slimming fashion" and "the more emotional lives of the younger generation since the War."[65] A 1926 article in *The New York Times* recounted the findings from a two-day conference on adult weight, held by the New York Academy of Sciences. Researchers at the conference reported an outbreak of eating problems that they linked to a "psychic contagion." One physician described a significant increase in the number of women whose dramatic weight reduction led to mental breakdown and hospitalization.[66]

However, "The flapper along with the entire exuberant culture of the 1920s vanished into the abyss of the Depression and then the consuming preoccupations of the Second World War."[67] Hemlines fell in the 1930s, and the defined waist returned. The ideal woman of the 1930s still had plenty of curves, but overall she remained slim.

The late 1940s and 1950s saw a temporary interruption of a long-term trend toward slenderness. Political and social reaction after World War II drove many white middle class women out of their war effort jobs and back to their kitchens.[68] One historian claimed that it was a period of "resurgent Victorianism."[69] As the economy switched back to making domestic products, it urged women to again focus on a consumer role. Young men used the G.I. Bill to pay for educations and buy first homes. The "family wage" was enough to support a family, and it also justified why women should be paid less. While women still went to college, their numbers in the professions declined and many opted to marry upon leaving school. It was a time of economic expansion, the rise of suburbia, and the spawning of the white middle-class housewife.[70]

To complete the picture of domestic bliss, American fashion in the 1950s revived the hourglass figure, created primarily by the girdle. Cinched waists, long, full skirts, and even crinolines came back to create a silhouette not unlike the Victorian lady. Hollywood provided a busty new feminine image, first personified by Marilyn Monroe, and later carried to extremes by the "Mammary Goddesses" like Jayne Mansfield.[71]

The Ultra-Slender Ideal

Within a decade, however, thin was back in. This time, the super-slim body ideal met and merged with other social influences. These forces included a new feminist movement and changes in women's roles, the increasing power of the media, and rampant consumerism. As Seid notes,

> The imperative to be thin became monolithic as fashion's decrees were reinforced and pushed by all cultural authorities—the health industry, the federal government, employers, teachers, religious leaders, and parents until the concept became so self sustaining, so internalized that no reinforcement was necessary.[72]

The women's movement of the 1960s offered alternative visions to the "happy housewife" of the 1950s. Women began to close the gap in higher education and the number of women in the labor force, with and without children, increased dramatically. The contraceptive revolution gave women some increased control over their own fertility.[73]

Yet as women gained economic, social, and political resources for charting their own destinies, the pressure to shrink in body size again returned.[74] The media began to play a dominant role in this pressure. In the 1960s, films were no longer the most important influence in defining beauty. Instead, television, the American fashion industry, and women's magazines became the arbiters of image. Fashion photography demanded stick-thin bodies that did not compete with the clothing.[75] In the mid-1960s, a 17-year old, 5-foot-6-inch model weighing 97 pounds entered the American fashion scene from England. Her name was Twiggy. She became an instant celebrity and many young women began to emulate her. Understandably, researchers point to this decade as the era of marked increase in eating disorders.[76] And, of course, another major fashion influence—the Barbie doll—had already arrived.

Writing in the magazine *Smithsonian*, Doug Stewart notes, "If all the Barbies sold since 1959 were laid head to heeled foot . . . they would circle the earth three and a half times."[77] Barbie demonstrates that while your roles can change over time, you may never find relief from the Cult of Thinness. "She was a model in 1959, a career girl in 1963, a surgeon in 1973, and an aerobics instructor in 1984,"[78] but her body dimensions have never changed. They include exaggerated breasts, impossibly long legs, nonexistent hips, and a waist tinier than a Victorian lady's. This is the perfect figure presented to little girls as "ideal."

Women's magazines, relying as they do on a consumer culture, also contribute to the obsession with image, fashion, and thinness. One researcher points out:

> . . . women's magazines collectively comprise a social institution which serves to foster and maintain a cult of femininity. This cult is manifested both as a social group to which all these born females can belong, and as a set of practices and beliefs: rites and rituals, sacrifices and ceremonies, whose periodic performance reaffirms a common femininity and shared group membership. In promoting a cult of femininity, these journals are not merely reflecting the female role in society; they are also supplying one source of definitions of, and socialisation into, that role.[79]

Why are women who have gained some economic independence still expressing their self reliance and inner control through these body rituals? An opposing view suggests that dieting and physical fitness are not methods for the subordination of women, but ways that women can feel powerful. After all, for many women, feeling fat means feeling powerless. However, by investing time, money, and energy on attaining a thin body, women may be substituting a momentary sense of power for "real authority." Some feminists take the argument even further, pointing out that being overweight is, itself, a way of expressing power. In *Fat Is a Feminist Issue*, Susie Orbach notes that being overweight is one way to say "no" to feeling powerless. A fat person defies Western notions of beauty, and challenges, in Orbach's words, "the ability of culture to turn women into mere products."[80] Kim Chernin says that in a feminist age, men feel drawn to and perhaps less threatened by women with childish bodies because "there is something less disturbing about the vulnerability and helplessness of a small child, and something truly disturbing about the body and mind of a mature woman."[81]

The fact remains that regardless of their economic worth, women are socialized to rely on their "natural" resources—beauty, charm, nurturance—to attract the opposite sex.[82] The stakes of physical attractiveness for women are high, since appearance, including body weight, affects social success.[83] Women are likely to experience even a few extra pounds as a major problem in their lives; they tend to weigh themselves frequently and report seeking medical help for weight problems more often than men. In fact, many women willingly embrace the mind/body dichotomy, partly because the woman who invests herself in her body often reaps enormous rewards and benefits. Ignoring investments in one's body can mean the loss of both self-esteem and social status.

Yet I do not want to leave the reader with the impression that all women are enslaved by bodily concerns (or that all men are the "enemy"). Throughout history, many have found ways to resist, or to alter the meaning of, the social practices controlling them. For example, the exaggeratedly corseted Victorian figure drew attention to the waist and enhanced the bosom. While the corset was initially intended as a means of feminine control, some women began to use it to express their sexuality. In time, political conservatives reacted to the way women subverted the intent of corseting. They decried the practice as a sign of loose morals.[84]

It would be hard to portray all women as the "victims." Women frequently collude in promoting body rituals. Like the mothers who bound their daughters' feet, or tightened their corsets, today's Mom may recommend the latest diet and fitness club to her daughter, or gently slap the girl's hand if she wants to have another piece of chocolate cake. Many women who are not in a position to change their basic social and economic environment may try to cut deals with the system. For Delia, in the previous chapter, dieting for a culturally correct body might help her catch a rich husband. For other young women, working out in the gym may build the confidence they need to compete with men in the work world.

In a way, we can consider women's bodies as cultural artifacts, continually molded by history and culture.[85] Subjected to such pressures, the "natural body" is lost. What replaces it may take the form of the bewigged eighteenth-century countess, the wasp-waisted Victorian housewife, the leggy flapper, or the waif modeling Calvin Klein jeans. All are bodily reflections of the play of power within a society.[86]

In the next chapter, we'll examine this power play in the "body business." The marriage between patriarchy and capitalism brings both institutions enormous rewards. Without looking at this alliance, we would lose an important piece of the Cult of Thinness puzzle.

3 There's No Business Like the Body Business: Food, Dieting, and Recovery

Mirror: Anything that faithfully reflects or gives a true picture of something else.[1]

"I lived with two girls for two years. And when they spoke to you they had to be standing in front of the mirror. One of my roommates finally said to me 'when we come in our room we're hanging a towel over the mirror.' Because they would literally stand there and look at themselves at every angle in the mirror the entire time. I'd say, 'What is your problem? Are you in love with yourselves, are you that concerned?' "—Judy, college senior

"It makes me feel better when I look in the mirror. And it shows me that I'm taking care of myself and it just makes me feel better about myself."—Juliet, college junior

"I would see these thin girls in the magazines and say I want to be like that. I would look at myself in the mirror and I didn't like what I saw."—Donna, college sophomore

"I always have to look in the mirror. It can get so frustrating because all mirrors are different, you know? I am always like, 'God, I wish I was ten pounds lighter.' "—Anne, college junior

The concept of a mirror gives us an analogy for how society fosters women's obsession with their weight and body image. A mirror reflects the virtual image of an object placed in front of it. The reflection Judy, Juliet, Donna, and Anne each observe in the mirror may be a true likeness, but their perceptions of this image are deeply distorted by their feelings about their bodies. Judy's roommates check the image obsessively; Juliet seeks in her reflection her own well-being; Donna fears she can't measure up; and Anne asserts that all mirrors differ. As we have seen in Chapter 2, our society encourages women to see themselves as objects.

In this chapter we can focus our own critical gaze on the mirror set up by capitalism and patriarchy. If, instead of looking at the bright reflective surface, we survey the greater structure framing and supporting this mirror, we are prompted to ask a different set of questions. Not "What can women do to meet the ideal?" but "Who benefits from women's excessive concern with thinness?" "How is this obsession created, promoted, and perpetuated?"

Profiting from Women's Bodies

Because women feel their bodies fail the beauty test, American industry benefits enormously, continually nurturing feminine insecurities. Ruling patriarchal interests, like corporate culture, the traditional family, the government, and the media also benefit. If women are so busy trying to control their bodies through dieting, excessive exercise, and self-improvement activities, they lose control over other important aspects of selfhood that might challenge the status quo.[2] In the words of one critic, "A secretary who bench-presses 150 pounds is still stuck in a dead-end job; a housewife who runs the marathon is still financially dependent on her husband."[3]

In creating women's concept of the ideal body image, the cultural mirror is more influential than the mirror reflecting peer group attitudes. Research has shown that women overestimate how thin a body their male and female peers desire. In a recent study using body silhouettes, college students of both sexes were asked to indicate an ideal female figure, the one that they believed most attractive to the same-sex peer and other-sex peer. Not only did the women select a thinner silhouette than the men,[4] but when asked to choose a *personal* ideal, rather than a peer ideal, the women selected an even skinnier model.

Advertisements and Beauty Advice: Buy, Try, Comply

Capitalism and patriarchy most often use the media to project the culturally desirable body to women. These images are everywhere—on TV, in the movies, on billboards, in print. Women's magazines, with their glossy pages of advertising, advertorials, and beauty advice, hold up an especially devious mirror. They offer to "help" to women, while presenting a standard nearly impossible to attain. As one college student named Nancy noted in our interviews,

> The advertisement showed me exactly what I should be, not what I was. I wasn't tall, I wasn't blonde, I wasn't skinny. I didn't have thin thighs, I didn't have a flat stomach. I am short, have brown curly hair, short legs. They did offer me solutions like dying my hair or a workout or the use of this cream to take away cellulite.

Women's beauty magazines such as *Vogue* and *Cosmopolitan* have always stressed the need for women to take an active interest in their looks. A 1957 article in *Vogue* entitled "How to Look Like a Beauty" cajoles women into taking action.

> Some women find beauty unnecessary, and will take the trouble to hide inherited good looks behind frowzy hair, fat, badly-chosen spectacles, and dreary clothes, feeling elemental and honest when they spurn artifice . . . But the woman who finds it necessary to be beautiful comes to look like a beauty because there is a need in her. She will make up for a lack of inherited good looks with work, knowledge, time, fashion and any artificial aid that's appropriate . . . beauty is, very often, like any other ambition or drive—capable of realization of a degree in proportion to the need of satisfaction . . . If her hair is a natural disaster, she goes to the very best hairdresser and gets the very best advice as well as the very best work that she can. She finds—just as she'd hoped—that the actual quality of the hair can be changed; that dyeing may give body to flyaway hair; that somewhere in chemistry there is the perfect permanent wave for her hair; that, surprisingly, a hair-straightener may be indicated.[5]

Marsha, a sophomore, gave the following account of where she got her drive to be thin.

> *Vogue!* *Vogue* magazine or anything like that. From society. I think *Vogue* is a good example of society and I think it puts a great deal of pressure on women everywhere. It's a terrible thing to be guilty of but if I'm with a friend and I see a woman that doesn't look good, I'll say something like 'Look at her hair. She should do something with it. God! She's heavy there.'

She also noted,

> "These magazines do make-overs. What can she do to make herself better? I think magazines are really mixed up. In the same magazine you'll see a health focus up-date and they'll write up things like . . . it's important for a woman to have a good perception of herself, not to worry too much about your looks and not to drive yourself into dieting . . . people will write about that stuff. But then you turn the page and there's Miss Jane who just had a make-over and this is what we did to her hair and face and now doesn't she look better, before and after pictures. Turn the page and there's an ad saying try our aging cream so you won't grow old tomorrow. So you find a mixture of things.

The "Beauty Helpline" section of *Cosmopolitan* in December 1992 contains typical advice. This month's question concerns what to do about "droopy upper eyelids." It poses as a helpful tip from a beauty editor, but is also a thinly disguised endorsement for a New York plastic surgeon.

> "QUESTION: I have droopy upper eyelids that give me a tired, puppy-dog look. How can I perk them up?
> ANSWER: Try these makeup tips. Start by tweezing brows so they end up

on an upward slant. Sweep neutral shadow across entire eye, then dust strips of medium tone across outer third of both crease and lashline (they should meet in a V). Use eye pencil on top lid only. Finish with two coats of lengthening mascara, concentrating application at center. If this technique proves inadequate, you might consider blepharoplasty, a simple outpatient plastic-surgery procedure. Area is numbed with local anesthesia, an incision is made across lid, then sagging matter (excess fat and skin) is removed, explains New York City plastic surgeon Cap Lesene. Expect to be black and blue for about four days and puffy for up to ten before new "lifted" look emerges. The cost runs from twenty-eight hundred to four thousand dollars.[6]

Not everyone is taken in, of course. One student I interviewed dismissed the images she saw in the advertising pages of magazines as "constructed people."

> I just stopped buying women's magazines. They are all telling you how to dress, how to look, what to wear, the type of clothes. And I think they are just ridiculous . . . You can take the most gorgeous model and make her look terrible. Just like you can take a person who is not that way and make them look beautiful. You can use airbrushing and many other techniques. These are not really people. They are constructed people.

Computer-enhanced photography has advanced far beyond the techniques that merely airbrushed blemishes, added highlights to hair, and lengthened the legs with a camera angle. The September 1994 issue of *Mirabella* featured as a cover model "an extraordinary image of great American beauty." According to the magazine, the photographer "hints that she's something of a split personality . . . it wasn't easy getting her together. Maybe her identity has something to do with the microchip floating through space, next to that gorgeous face . . . true American beauty is a combination of elements from all over the world." In other words, the photo is a computerized composite. It is interesting that Mirabella's "melting pot" American beauty has white skin and predominantly Caucasian features, with just a hint of other ethnicities.

There are a number of industries that help to promote image, weight, and body obsession, especially among women. If we examine the American food and weight loss industries, we'll understand how their corporate practices and advertising campaigns perpetuate the American woman's dissatisfaction with her looks.

The American Food Industry: Fatten Up and Slim Down

"I pop back and forth, gain and lose, gain and lose. It's so weird because you can see these pictures of me in December, my face will be thin, and by January it'll plump up again. For example, last Christmas my mother put me on a strict diet.

I dropped about ten pounds. When I got back to school, I just ballooned as soon as I got here, the same old routine as before." Stephanie, a sophomore, knows the yo-yo syndrome well. So do millions of other Americans.

Obesity specialist Dr. Thomas Wadden complains that for every dollar spent by the Surgeon General and researchers to prevent or treat obesity, the food industry spends one hundred dollars to get people to buy junk food. "We're being fattened up by the food industry and slimmed down by the twelve-billion-dollar diet and exercise industry. That's great for the capitalist system, but it's not so great for the consumer."[7]

It is not uncommon for the average American to have a diet cola in one hand and high-fat fries and a burger in the other. Food and weight loss are inescapably a key part of the culture of the 1990's. The media bombard us with images of every imaginable type of food—snack foods, fast foods, gourmet foods, health foods, and junk foods. Most of these messages target children, who are very impressionable, and women, who make the purchasing decisions for themselves and their families. At the same time women are subjected to an onslaught of articles, books, videos, tapes, and TV talk shows devoted to dieting and the maintenance of sleek and supple figures. The conflicting images of pleasurable consumption and an ever leaner body type give us a food consciousness loaded with tension and ambivalence.

Social psychologist Brett Silverstein explains that the food industry, like all industries under capitalism, is always striving to maximize profit, growth, concentration, and control. It does so at the expense of the food consumer. "[It] promotes snacking so that consumers will have more than three opportunities a day to consume food, replaces free water with purchased soft drinks, presents desserts as the ultimate reward, and bombards women and children with artificially glamorized images of highly processed foods."[8]

Diet foods are an especially profitable segment of the business. A few selected statistics follow:

- In 1980, diet foods equaled roughly 7 percent of all U.S. food sales.[9]

- By 1981, more than eight million pounds of saccharin were being produced in the United States for a $2 billion domestic market.[10]

- In 1984, sales of diet food and beverages were advancing at three times the pace of the rest of the food market.[11]

- By late 1990, the diet/lo-cal segment of the frozen entree category alone was worth $689 million.[12]

In 1983, the food industry came up with a brilliant marketing concept, and introduced 91 new "lite" fat-reduced or calorie-reduced foods.[13] The success of lite products has been phenomenal. The consumer equated "lightness"

Figure 2 *Featherlite by Subtract,® September, 1982*

with health. The food industry seemed to equate it with their own expenses—lite foods have lower production costs than "regular" lines, but they are often priced higher:

"Light beer had more water. (Coors, one model for Miller Lite, was so watery that detractors called it 'Colorado Kool-Aid') . . . The saccharin in diet drinks was cheaper than sugar . . . Especially profitable were dairy analogs fabricated from inexpensive vegetable fats: margarine, egg substitutes, creamers, frozen dinners, 'lite' cheeses."[14] The Food and Drug Administration has since tightened its regulations on food labeling and what "lite" claims are made.

Manufacturers have reaped uneven profits in recent years. In 1992, diet and low fat dinners and entrees alone constituted a $3.3 billion market.[15] In that year, sales for Kraft Free salad dressing grew 9% over the previous year, while sales of Miracle Whip Free went down 11%. For the same year Nabisco could not keep up with market demand for SnackWell's Devil's Food Cookie Cakes, while Ralston Purina experienced a 29% decline in sales for Hostess Lights snack cakes.[16] In 1992, ConAgra budgeted $200 million for marketing its Healthy Choice line of frozen dinners and entrees, with $50 million of that amount dedicated to media advertising. Projected sales for all Healthy Choice products for that year were $1 billion. For the year ending May 1995, sales of Healthy Choice frozen dinners and entrees were $550 million, representing the highest industry market share. Its spaghetti sauce alone posted $23 million in sales in its first six months on the market.[17] Price erosion, stemming from intense price competition, as well as inconsistent customer demand, has hampered the reliability of profits. This has caused manufacturers to regroup and rethink their strategies, eliminating some products, while expanding in other areas. This market is clearly not being abandoned, just reconsidered. For example, Weight Watchers has allocated more than $10 million for a new marketing campaign where for every 20 Weight Watchers products purchased, the consumer receives a $10 voucher toward future purchases.[18]

Yet with all this focus on eating well and eating "light," the American population is getting fatter. A Prevention Index Study by Louis Harris and Associates, cited in *American Demographics* magazine, indicates that in 1983, 58% of the adult population age 25 years and older weighed more than recommended for their height and frame size. In 1987, the figure rose slightly to 59%, climbing to 63% in 1992.[19] According to a 1994 study reported in the *Journal of the American Medical Association*, "The percentage of Americans who are obese has gone from one quarter of the population to one-third, and that's a huge increase." [20]

In creating the consumer demand for diet and junk fare at the same time, the food industry has favored mixed marriages and strange bedfellows. For example, ConAgra (best known for its line of frozen Banquet dinners) sells Morton, Chun King, and (through Beatrice) Clark Bars, La Choy, and Rosarita products, none particularly health-promoting. However, ConAgra also introduced, through Beatrice, the Healthy Choice reduced calorie frozen dinners.[21] Moreover, ConAgra has teamed up with Thompson Pharmaceuticals to sell Ultra Slim-Fast, an extremely popular liquid diet.[22] Eaten too many Clark bars and Morton Frozen dinners? Two meals a day of Ultra Slim-

Fast and one Healthy Choice frozen dinner will help you lose weight and help ConAgra make more money.[23]

And the ironies continue. Nestle, maker of dozens of chocolate and ice cream products, also makes Stouffer's Lean Cuisine, Stouffer's Right Course, and Carnation Slender products. Louis Sherry, maker of sugarless jams, also makes super rich ice cream;[24] General Mills offers Yoplait yogurt for the days when their Count Chocula, Lucky Charms, and Trix cereals are just too sugary.

With the latest rounds of corporate consolidation, where one food company expands by purchasing smaller companies, the opportunities to sell us diet and junk food simultaneously have multiplied.

Who Controls the Dinner Table?

"Behind the phantasmagoria of food brand names presented on television and in modern supermarkets, there are fewer and fewer processors and marketers gaining more and more control over dinner,"[25] says Jim Hightower, author of *Food Profiteering in America*.

Economist Lois Therrien explains, "The economic imperative driving consolidation is the search for growth. The volume of food sold in the U.S. is rising by only about 1 percent a year. Combine this with the skyrocketing cost of rolling out new products, and it's easier, and cheaper, to buy market share than build it."[26]

The connection between consolidation and profit is quite clear. In 1977, four companies—Kellogg's, General Mills, General Foods, and Quaker—produced 90% of the cereal sold in America,[27] a proportion that does not seem to have changed dramatically since then. In 1989, Kellogg's' net income was $465 million, General Mills' was over $300 million, and Quaker's was over $200 million. General Foods had been absorbed by Kraft—which had been taken over by Phillip Morris, whose 1989 net earnings were $2,425 million.[28]

Nor are the takeovers limited to food corporations. RJR Tobacco had already taken over Nabisco for $4.9 billion when it was taken over by Kohlberg Kravis Roberts & Company for $24.5 billion.[29]

Given the cultural pressures on women to be thin, it is no wonder that disordered eating develops in such an environment. Profit-building food juggernauts and their billion-dollar "fatten up/slim down" ad budgets have come up with some pretty powerful examples of schizoid thinking. In one of my favorite ads, the Hershey company invites us to savor the buttery toffee and chocolate of the wafer-like Skor candy bar. It shows the ultra-thin bar in profile, its wrapper seductively pulled back. The copy reminds us that even a candy bar "can never be too rich or too thin."

The Diet and Weight-Loss Industry: We'll Show You the Way

In October 1992, *Working Woman* magazine reported that at any given time 65 million Americans are dieting, spending more than $30 billion annually in the pursuit of losing weight.[30]

Increasingly, American women are told that they can have the right body if only they consume more and more products. They can change the color of their eyes with tinted contacts, they can have a tanned skin by using self-tanning lotion. They can buy cellulite control cream, spot firming cream, even contouring shower and bath firming gel to get rid of the "dimpled" look. One diet capsule on the market is supposed to be the "fat cure." It is called Anorex-eck, evoking the sometimes fatal eating disorder known as anorexia. It promises to "eliminate the cause of fat formation . . . so quickly and so effectively you will know from the very start why it has taken more than 15 years of research . . . to finally bring you . . . an ultimate cure for fat!"[31]

Many women believe that in order to lose weight they need to *buy* something, whether it be a pill, a food plan, or membership in a self-help group. The journey to creating the right body often begins with the purchase of a diet book. Today, some best-selling books include *The T-Factor Diet: Lose Weight Safely and Quickly Without Cutting Calories or Even Counting Them* (a #1 *New York Times* bestseller with two million copies sold), *The Last Ten Pounds, Stop the Insanity, Eating Leaner and Lighter*, and *Outsmarting the Female Fat Cell*. True adherents of the Cult of Thinness consider their diet books as bibles. Each book trumpets the true path to health and happiness, and holds up a mirror to reflect only its own narrow world of correct behavior. Some books open with "sermons" preached to the dieter. Others prescribe certain daily rituals of food preparation, food combination, food measuring, eating, and weighing oneself. Dieters are warned that they must follow the recipes carefully and must never "sin" by going off the diet.

If following the diet book doesn't work, many find solace in more formalized diet programs. "In 1991, 7.9 million people enrolled in commercial weight-loss programs generating more than $2 billion in revenue for these plans," notes John LaRosa of Market Data Enterprises.[32]

There are currently more than 17,000 different diet plans, products, and programs from which to choose.[33] Typically, these plans are geared to the female market. They are loaded with promises of quick weight loss and delicious low-calorie meals. Three of the most popular programs are Jenny Craig, the Diet Center, and Nutri/System.[34] Jenny Craig is the founder of a successful weight loss business with over 400 centers. Her annual revenues are more than $400 million. Memories of her childhood motivated Craig to start thinking about ways of losing weight for herself and others. "I used to look in the mirror and cry," she says, recalling her pudgy young reflection. "I would just cry and say, 'What did you do to yourself?' "[35]

Testimonials and before-and-after pictures are the primary marketing methods. "Our client Terry lost 92 lbs.," says the Nutri-Systems brochure. We see a "before" picture of Terry, overweight and slouching, clasping her hands together. Next we see Terry, 92 pounds thinner. She stands tall with her hands on her hips (she has a waist). She is wearing a leotard and sneakers. In the same brochure, we also meet the smiling Cheryl who has lost 116 pounds. There are more happy faces of women who have lost weight and who are now able to do things such as bicycling or even going out on a date with a handsome man. The promise is "a happier, healthier life."

The typical weight loss plan costs about $500 to $900,[36] though one of the best-known diet programs, Weight Watchers, is considerably less. Like its many clones, Weight Watchers is loosely run and mostly consists of a weigh-in, classes on nutrition, tips on handling various diet crises, and group support.[37] The initial registration fee is $14 to $17, with weekly meeting fees of $10 to $13.[38] The claimed weight loss per week is one to two pounds.[39] Since its inception in 1963, Weight Watchers has had more than 25 million members.[40] In 1988, an average of one million people attended Weight Watchers classes in 24 countries,[41] spending an average of $5 million in fees.[42] That same year, its owner, Heinz, sold $780 million worth of Weight Watchers brand food. Roughly 60% of Weight Watchers' annual worldwide sales of over $1.3 billion has been from low-calorie food, the rest from franchises and publications.[43]

Comprehensive diet centers with more personally supervised medical and nutritional attention are also popular. The Ferguson Diet Center had 2,300 locations in the United States and Canada in 1988, and earnings of $8.9 million.[44] Among the most highly supervised programs are the Very Low Calorie Diets such as Optifast, Medifast, or Health Management Resources. They are usually run by hospitals on an outpatient basis, with a standard regimen costing $2,000 to $5,000.[45] By serving "300 dieters annually, a typical hospital could increase its revenues by more than $800,000 a year."[46] The hospital-based weight-loss market was estimated at $5.49 billion in 1989.[47] One such program, United Weight Control, costs $2,500 for a 17-week liquid diet and $350 for a 26-week weight maintenance program. United Weight Control has treated 4,000 people since it was started in 1986.[48] Very Low Calorie Diets usually consist of powdered drinks that are supposed to completely substitute for meals. There have been a number of deaths attributed to them [49] and it is generally agreed that they carry some medical risks.

Thin Promises

Many of these programs promise weight reduction that is not only fast, but permanent. For instance, the brochure from the Jenny Craig Weight Loss Center reads:

> We listen to you. We work with you. We care about you and your success . . . that's what makes the Jenny Craig Program so unique. And its results are so long lasting. You see, my program does more than help you lose weight quickly and easily. It teaches you how to keep your weight off.

The Diet Workshop promises an even speedier loss if you join one of its Quick Loss Clinics. By making a firm six-week commitment, the dieter is rewarded by a loss of up to 20 pounds.

Yet there is virtually no data available to support the weight loss industries' claims. In fact, it has recently been noted that "many people who complete a commercial diet will regain one-third of their lost weight after one year, two-thirds or more after three years and most, if not all, in three to five years. Many do not complete the program."[50] Medical research has come up with a theory that "body weight may be regulated by a biologic set point system in the hypothalamus that 'defends' a particular weight by maintaining energy balance and food consumption at certain levels . . . traditional diets often fail to reduce weight because the lower food intake sets off a 'starvation reaction,' in which the basal metabolic rate and overall activity level are decreased." This research indicates that to lose weight (1) you need to increase the basal metabolic rate (aerobic exercise is one way) and (2) you must change the composition of your diet rather than restrict overall intake.[51]

Some of the women I interviewed echoed these observations.

> When I was really heavy at 17, I actually joined Weight Watchers. I lost eight pounds with them. Then I just decided I didn't like that structure, weigh your food and this and that. I came off and I was caught up in this new job and getting ready to go away to school. I ended up losing more weight. And then again six or seven months ago I joined some diet group. I figured it would be an incentive getting weighed once a week. And I did lose ten pounds relatively quickly and consciously. But recently I put it back on.

Another told me: "In high school I tried the Scarsdale diet. It worked to a point but then I decided I just lost my energy instead of losing weight. So now I really believe that there is probably a certain weight that I will never go below. Maybe I could be five to ten pounds less than I am now, but I would have to change my whole eating habits, reduce portions or give up drinking. So that might happen, but I don't do any of these trick diets any more."

Many of these diets require the participant to fixate on food to a great extent, creating an obsession with eating. One Vermont company called Weight Wizards produces a "foodmeter," which shows graphically how many calories are consumed for breakfast, lunch, dinner, and snacks. There is even a "color me thin weight loss kit." Many of the women I talked with mentioned this fixation.

> My little sister went on a diet, and she wanted me to go on one with her, so I said OK, I would this one week. I didn't really want to lose the weight. I just said I'll do it because she just wanted some moral support. After a week I said you're on your own. I never would have eaten this little. I became so hungry. How can I live on just pieces of toast? I never thought about food so much before I went on this diet. Whenever I'm hungry I eat. And if I'm not, I don't."

Julia joined a weight-loss clinic program and became more and more obsessed with losing weight: "It's just a starvation diet and they give you vitamins and you go in everyday and weigh. The focus was crazy. It was extreme. The focus was on Julia losing weight for the next three months . Fine, I could lose the weight, but then now that I've gained the weight back what does that mean? You put all this energy into losing weight, now that you've gained it back does that mean you're a screw up?"

Many of these programs produce food products that they encourage the dieter to buy. The Jenny Craig member receives a set of pre-packaged meals that cost about $10 per day. (It allows for some outside food as well.) Some diet companies are concerned with the problem of gaining weight back and have developed "maintenance" products. Maintenance programs are often expensive and their long-term outcomes are unproven. What *can* be proven are bigger profits and longer dependence on their programs.

The Dis-eased Body: Medicalizing Women's Body Issues

The therapeutic and medical communities tend to categorize women's eating and weight problems as a disease.[52] In this view, behavior like self-starvation or compulsive eating is often called an addiction. An addiction model of behavior assumes that the cause and the cure of the problem lies within the individual. Such an emphasis fails to examine the larger mirrors that society holds up to the individual.[53]

The Recovery and Self-Help Markets

One important individual therapeutic index can be seen in the variety of self-help books that have flooded the marketplace, reaping millions of dollars for the recovery industry. Recent estimates place over 15 million Americans in 500,000 recovery groups. Recovery books are booming and women are responsible for 75 to 85% of their total sales. Thomas Nelson's *Love Hunger* sold more than 200,000 copies in 1990, its first year of publication. As of May 1995, Melody Beattie's *Codependent No More* had sold five million copies. In its first six months, Anne Wilson Schaef's *Meditations for Women Who Do Too Much* had sold 10,000 to 20,000 copies a month. Harper/Hazelden, one of the major recovery book lines, put out its first meditation manual, *24 Hours a Day*, in 1952, which has since sold seven million copies.[54] Hazelden listed approximately 80 books at the beginning of 1990,[55] including at least eight on compulsive eating, published during the 1980s. Hazelden's mail-order business was selling to 98 countries, with roughly 1,500 products in their catalog.[56]

The president of Health Communications Incorporated, another major recovery book publisher, predicted sales of over three million books in 1990.[57] As of the end of 1989, HCI had produced 30 books a year, and now claims to have 145 titles. One of HCI's 1990 books was *On Shame and Body Image: Culture and the Compulsive Eater*. Comp Care, the third major recovery book press, claims to have approximately 1,000 titles, including several on food. One book of meditations, called *The Thin Book*, dates back to 1978.[58]

While a disease model lessens the burden of guilt and shame and may free people to work on change, it also has political significance. According to feminist theorist Bette S. Tallen, "The reality of oppression is replaced with the metaphor of addiction." It places the problem's cause within a biological realm, away from outside social forces.[59] Issues such as poverty, lack of education and opportunity, racial and gender inequality remain unexamined. More important, a disease-oriented model of addiction, involving treatment by the health care system, results in profits for the medical-industrial complex. Addiction, Tallen notes, suggests a solution that is personal—"Get treatment!"—rather than political—"Smash patriarchy!" It replaces the feminist view, that the personal is political, with the attitude of "therapism," that the "political is personal."[60] One of Bette Tallen's students told her that she had learned a lot from reading *Women Who Love Too Much* after her divorce from a man who had beat her. Tallen suggested that "perhaps the best book to read would not be about women who love too much but about men who hit too much."[61]

The idea that overweight is a disease, and overeating represents an addiction, reinforces the dis-ease that American women feel about their bodies. The capitalist and patriarchal mirror held before them supports and maintains their obsession and insecurity. We have seen how the food, diet, and recovery industries "feed" on this insecurity. In the next chapter, we'll look at how the fitness and cosmetic surgery businesses benefit as well.

4 There's No Business Like the Body Business: Fitness and Cosmetic Surgery

"I want you to watch the mirror like a hawk! Make sure you are squeezing your muscles in your thighs tight the whole time! Now I want you to work on your abs. Pretend that you have on a steel girdle. You are going to tighten everything and not let up! In order to tighten your abdominal muscles pretend that your belly button is touching your spinal column. Squeeze! Are you squeezing?"—Annette, fitness trainer

The Right Body: Sculpting with Exercise

The words in vogue at my fitness club are "tight" and "steel," as in "buns of steel" and "abs of steel." "Discipline," "strength," and "power" are also part of the vocabulary. "Muscle definition" is the term used by trainers to urge both men and women to sculpt their bodies. To make the point, the club has changed the name of my aerobics and conditioning class to "Body Defined." Some of the women here think of their different body parts as fashion accessories. One works on the triceps so that she can show off her arms in a new sleeveless shirt, another counts situps in order to reveal her "abs of steel" in a bikini.

Our cultural mirrors have undergone a massive transition from the 1950s, when the ideal female body had soft curves. The illustrated pages of *Cosmopolitan* and *Vogue* reflect these changes over the past four decades. In the late 1950s, the tiny waistline was in. Clothing was tapered to fit, emphasizing the shoulders, waist, and hips. Women relied on girdles to achieve this look.

During the 1960s fashion moved away from the hourglass shape to a more stick-like figure. The waistline disappeared. Girdles, while still important aids to the slim figure, were becoming lighter and more flexible. The real move toward slimness became very pronounced later in the decade. The mini skirt shifted attention to the legs, shaped by pantyhose.

Today, a woman's body is supposed to be in shape on its own, liberated from the girdle. Women are expected to be thin and firm with exercise and dieting. A more recent ad for foundation garments promotes the ease and freedom of wearing light, stretchy fabrics. All a woman needs is the "light

37 minutes. 49 minutes. 39 minutes. 44 minutes.

Figure 3 Firm Parts™ Workouts, December, 1994

firmness of Lycra® spandex"—she shapes her underwear, rather than the other way around.

Ironically, the emphasis on a firm, shapely body has created another set of demands. This toned, "liberated" body takes a great deal of time and energy to create.

The Marketing of the Slender Amazon

The interest in physical fitness for both men and women has been growing since the 1970s. Getting in shape with exercise is now considered an essential part of a healthful lifestyle.[1] and the 1980s brought a more muscular ideal of the female body. The subcutaneous fat layer, which gives softness to the female physique, disappeared in favor of large, hard muscles. *Time* magazine devoted a cover story to this entitled "New Ideal of Beauty."[2] Examples included actress/fitness gurus like Jane Fonda and Victoria Principal.[3]

Annette, the fitness trainer quoted above, commented:

> Women come to me and tell me they want their bodies changed . . . they want everything smaller, tighter. They want 'cuts,' like in male bodybuilding, where you can see the muscle right under the skin, no fat anywhere, the very thing that's feminine. But it's hard for a woman to develop cuts. It requires extreme leanness and a good amount of testosterone, which most women just don't have much of in their bodies.

Early religious asceticism involved the acceptance of persecutions. Believers denied their physical needs for sleep or food, and so on, as a means for the "expiation of sin, self-conquest, the intercession for divine graces and favours, and the imitation of Christ."[4] The fitness movement can be seen as a new form of asceticism, which "serves no higher moral aim but which, paradoxically, promises pleasure through self-denial."[5] Fitness fanatics follow

Figure 4 Tapertop by Jantzen,® September, 1957

specific painful rituals, often compulsively.[6] Annette mentioned her clients who never miss their daily workouts, who pay careful attention to the "correct" costume, who insist on certain spots in class with the best mirror view. They worship certain aerobics instructors, or are attached to a particular fitness machine. As Annette put it, "One club member with the flu literally crawled down her apartment stairs to come and try to use the Stairmaster . . . People are in bondage—it's not something they do, the thing does them." These rituals have proved to be a boon for the fitness industry.

Figure 5 TWA, and STJ, Up Up and Away, 1967.

Fitness has become a $42.9 billion industry, encompassing health clubs, exercise videos, home exercise equipment, clothing, and accessories.[7] Who constitutes the fitness market? According to the Lifestyle Market Analyst 1993, the median consumer's age is 40.8 years. Fifty-three percent are between 25 and 44 years; 54% are married, 22.2% are single males, and 23.7% single females. The median household income is $38,139.[8]

Just twelve years ago, the Census of Service Industries lumped "fitness clubs" together with sports and athletic clubs.[9] Today it is a distinct category.

Figure 6 The Liberated Body by Lily of France,™ March, 1985

Similarly, the Census of Manufacturing has separated "home fitness equipment" from playground and gymnasium gear. Exercise as play; exercise as sport; exercise as weight loss technique—the same activity can have different meanings and different intents.

The U.S. Industrial Outlook 1993 indicates that "the exercise and fitness sector of the sporting goods industry has been the fastest growing since the end of the 1980s . . . exercising with equipment was the seventh most popular activity in terms of participation."[10] Since 1980, sales volume has tripled, with Americans buying $1.79 billion worth of home fitness equipment in

1990.[11] NordicTrack leads the way, making a profit of $83 million on sales of $267 million in 1992[12]—an increase of more than 500% since 1986.[13]

Americans are also joining health clubs. One industry expert estimates that 30 million people belong to these clubs, up from 20 million five years ago. And these figures do not include those who work out in corporate or residential fitness facilities or are members of nonprofit clubs.[14]

Exercise videos have been a burgeoning market since Jane Fonda pioneered the genre in 1982 (she has since sold over 8.5 million videos).[15] Consumers spent $265 million on these exercise tapes in 1992, $285 million in 1993, and are projected to spend $290 million in 1994.[16] In the spirit of expanding the market, many video companies offer "mini" exercise programs with "modular sections that can be completed in about a half-hour" since "consumers' schedules are usually tighter than their buttocks."[17] The market continues to be segmented, with new emphasis on toning and shaping as distinct from aerobics.

The health and fitness industry appears in business literature as a well-established market. It plays the same "game" as any other industry. It uses the same tools of market and consumer analysis, offering products in pursuit of greater market shares. A stream of heavily marketed new products, like the NordicTrack executive exercise chair or Jane Fonda's latest "favorites" video, created big profits and drives the industry.

A video trade journal states, "Getting a toe-hold in this market requires creative strategies that run the gamut from no-cost marketing—reviews in magazines—to megabuck campaigns." And a marketing executive says, "It's clear we spend a lot of dollars on advertising . . . We spent a couple of million [on] the *Dancin' Grannies* tape in direct consumer advertising. It's one of the major reasons why our programs are so successful."[18]

NordicTrack has opened a chain of outlets in shopping malls, concerned that "it has been missing out on half its potential market: people who are reluctant to buy a $600 exerciser over the telephone without trying it first." These stores carry no stock, only demonstration models. They are extremely profitable, earning more than $1,400 per square foot.[19]

While equipment salespeople, fitness club managers, and video instructors may be speaking the language of concern for one's personal health and well-being, the industry literature clearly speaks the language of business. As Annette said of the club where she works: "The problem is that the people who own the clubs are part of the fitness cult themselves—they subscribe to a lot of the behaviors as the members. So they're not necessarily going to be promoting moderation when they are part of this fanaticism. And it's money! If you have somebody coming in here and paying for four aerobics classes a day you are making money off that person, even if she can barely crawl out afterwards. But . . . you can't give people something they don't want."

And there are signs that what people want in terms of fitness may change. A large segment of the American population, the baby boomers, is in fact getting older and fatter.[20] An article in *American Demographics* points out, "In 1983, 58% of adults weighed more than is recommended for their height and frame size. That share now stands at 63% according to the 1992 *Prevention Index* published by *Prevention* magazine . . . Aging inevitably changes a body's shape, and aging baby boomers usually have expanding waistlines. Yet the boomers' strong attachment to youth means that their anxiety increases with their weight. Advertisers make matters worse by presenting images of thin models to their plump customers. In the 1990s businesses will be rewarded if they can show boomers healthy ways to surrender to the bulge."[21]

There is a huge gap between the reality of our own bodies and the images to which we aspire to. Yet we continue to go for the image. And business gladly serves us in this chase as long as it continues to be profitable. The American *Demographics* article hints that there is a limit to how much longer the large baby boomer market segment will support this game as they live with the reality of a physically aging body. We will see how the players in the fitness industry respond.

The Right Body: Sculpting with Surgery

A fundamental change is taking place in American medicine. The increased corporate ownership of hospital chains and physician groups has been termed a "medical-industrial complex."[22] Some researchers point to the rise in the number of for-profit hospital chains that are claiming a growing share of the medical market.[23] In many ways the medical establishment has become a capitalist system of production that is subject to "a contradiction between the pursuit of health and the pursuit of profit."[24]

The medical industry needs revenue for underused facilities, and very often the solution is to drum up some business geared to women's body insecurity. As one analyst notes,

> Pick up a magazine that carries glossy ads, and you will see a full-color photograph of a wistful-looking woman in her thirties, wearing dancing tights. Her overall appearance lends a palpable persuasiveness to the copy that invites her to look carefully at her body—her falling buttocks, her flabby thighs, her sagging breasts, her aging face—and to do something about it. The ad is copyrighted by one of the large hospital corporations, Humana Inc. The corporation, it would seem, is seeking more revenue from one of its underutilized operating rooms by playing, as ads have always played, on the weaknesses and insecurities of frail humanity. But this time it's not a cosmetic or a mouthwash that's being hawked—it's invasive surgery that, even under conditions of necessity, should

not be lightly undertaken. But corporate practice calls for each branch of the "operation" (the appropriate word here) to earn its share of profits. Staff surgeons must cut; hospital beds must be filled. Besides, who can really hold a corporation responsible for what an individual freely chooses to do? The surgery, when it takes place, will in every way be voluntary."[25]

Plastic surgery has become a $5-billion-a-year industry and is increasingly considered part of the natural order of things for women.[26] In fact, some research has shown that such surgery has come to be viewed as a "moral imperative"—an "extension of a woman's regard for her appearance and therefore an expression of her essential femininity."[27]

Rhinoplasty, the ubiquitous "nose job," is one of the most popular plastic surgery procedures. Jacques Joseph (in Berlin during the 1920s and 1930s) was the first plastic surgeon to suggest that "abnormal appearance can scar the personality."[28] He developed and advanced the art of shaping noses. Today rhinoplasty is an important, if painful, beauty ritual for many young women.

When I walked into a plastic surgeon's office to interview him, I overheard a conversation between a mother and daughter. The daughter was around 18 years old and her mother had brought her in for a consultation.

MOTHER: "As your mother, I love you and I want you to have every opportunity to be the best that you can be. I feel that you will benefit from having your nose done. I'm bringing you in here, but it's clear to me that you don't want it."

DAUGHTER: "I want people to like me for the way I am! If I have to do this to catch a future husband, then I'll do it, but I think they should like me the way I am."

After a long wait, the doctor called them in. Sitting in the empty waiting room, I could hear the angry voices of mother and daughter. When they left, I met with the surgeon for our interview. I told him I could not help overhearing the conversation from the waiting area and asked him how things got resolved. Without revealing any confidential information, he replied that the daughter was not getting a nose job at this time. He mentioned that the mother had mixed emotions and that while she was very proud of her daughter for saying "Let people love me for what I am," she also felt that in time her daughter "will realize that this is a process of self-improvement. Just like improving your mind and character, doing something to look better may be seen in the same vein." The surgeon did indicate that he would never push anyone into surgery, but felt that eventually the daughter would most likely return.

Surgery as "self-improvement" is increasingly becoming an option, even a mandate. It is often fueled by the plastic surgery industry itself. And the

Figure 7   Beauty and Fitness Cosmetic Center of Beverly Hills, 1993

market has expanded, as one recent survey pointed out, with nearly one in three patients having incomes of less than $25,000 a year.[29]

Another plastic surgeon I talked with expanded on how medicine has become a growing business. It is finding high profits in trimming fat and removing wrinkles from women's bodies.

Over the last decade cosmetic surgery has been growing by leaps and bounds. Plastic surgery is now not only for the upper class but the middle class as well. Competitiveness in the medical marketplace has made these procedures within reach of even the lower middle class. In New York they now have a chain of clinics which specialize in cosmetic surgery marketed to the masses—assembly line cosmetic surgery. Those who own the operation are not necessarily the doctors. Some surgeons feel pressured to drop their standards because they are economically driven by competitive forces. For example, in Massachusetts an M.D. is licensed to perform any type of plastic surgery he or she wants. If they don't want to spend the money opening up a surgery clinic all they need to do is to team up with an MBA who buys a clinic and says to the doctor, "OK, get to work and start doing suction lipectomies. Do as many as you can." There is no law that says he can't. He hires an M.D. with no training and says, "Get in there, we'll show you a movie on how to do it. I'll pay you $500 per person, and I'll charge $1,000 for plastic surgery. More certified plastic surgeons are charging $2,000. I'll undercut them by half." So they've stolen the soul of medicine.

According to a survey by the American Society of Plastic and Reconstructive Surgeons, approximately 94% of their patients are women. One of the most popular procedures performed by surgeons to help women slim down is the procedure known as liposuction.

Liposuction, or the scooping out of body fat, became the number one aesthetic surgical procedure in America in 1986. The 99,330 cases that year showed an increase of almost 50% since 1984. Fees ranged from $500 to $4,000. (There were also 32,340 abdominoplasty or "tummy-tuck" surgeries done, at $2,000 to $6,000 apiece.) Cosmetic operations in general increased 24% from 1984 to 1986. Today, liposuction is the third most common procedure.[30] Currently, the Liposuction Institute of Boston charges $95 for a consultation and $4,000 to $7,000 for the surgery performed by a Board Certified physician.

The process of liposuction has been described as follows:

Through an incision no more than a half-inch long . . . the doctor inserts the hollow steel cannula, which has several small perforations on one side of the tip, and moves it around just under the skin. The smooth tip pushes most of the tough blood vessels and nerves aside, while a vacuum extractor pulls the softer fat in through the side holes. By artful manipulation of the cannula, the doctor recontours the fat.[31]

Breast augmentation has also been a popular procedure. Between one and two million women in the United States alone have had implants.[32] Only 20% of these women submitted to this procedure as part of a postmastectomy; 80% of breast reconstructive surgeries are done on healthy women who want to change their breast size. This may reflect the obsession with thinness in another way. Women who strive for ultra lean figures often find that their breast fat disappears. Increasingly, the "right" female body is an amalgam of the impossible, demanding flat stomachs, thin thighs, and boyish hips, yet large breasts. Silicone breast enlargement once accounted for $450 million a year, a seemingly endless market served by eager surgeons. Dr. Adriane Fugh-

Berman, medical advisor to the National Women's Health Network, wrote, "In the early 1980s the American Society of Plastic and Reconstructive Surgeons reached a new low when it suggested that small breasts be considered a disease—they named it 'micromastia.'"[33] But there is mounting evidence that the cure for this "disease" has backfired. A class-action lawsuit brought by women claiming their medical problems were caused by their silicone implants resulted in a $3.7 billion settlement by Dow Corning, Bristol-Myers Squibb Company and Baxter Healthcare in 1994. To date, about 16,000 women have filed suit in federal and state courts, claiming injury. The companies involved said that they agreed to the settlement, which took two years to negotiate, in order to get the litigation behind them. They insisted that the implants were safe despite the wide range of immune-related problems reported by the women.[34]

Though relatively recent, these liposuction and implant technologies are simple to execute and can be done quickly . They are also lucrative. As one surgeon I interviewed pointed out,

> In the past twenty years the technological advancements in plastic surgery have allowed people to alter any part of their body within a few hours, and to literally buy a beautiful face or body. Liposuction, for example, allows you to accomplish something that no human being could ever do before: to chose where to lose weight. Instead of going to a health club and working out for one year, and still not being satisfied, you can point to a place and say, 'Doctor, I want to get rid of this specifically.' We've never had that luxury of choice before. You can pretty much dictate a type of body that you want, though there still are certain shapes that are intrinsic to a person and cannot be created."

Some surgeons are perfecting the method of "fat grafting" whereby fat is suctioned from one part of the body, for example, the thigh, and injected into another part, such as the cheek or lip area in order to fill in a wrinkle.[35]

There are even more drastic surgical lengths to which people will go to lose weight. Those considered morbidly obese (160 to 225% above ideal body weight) with no hope of losing weight can choose "stomach stapling." In this procedure, "the stomach is reduced in size by applying four rows of stainless steel staples across the top of the stomach. An opening is made in the upper pouch of the stomach, and a portion of the small intestine is attached to this opening."[36]

In 1983 the cost of gastric stapling ran $3,000 to $10,000.[37] Today, Boston's Brigham and Women's Hospital charges roughly $3,500 dollars for the procedure, and New England Deaconness Hospital charges $3,800 for the physician's fee alone.

I interviewed one woman who had her stomach stapled as a last resort, believing that it would change her life when nothing else had worked.

Janet's Story

When Janet went to nursing school she weighed about 185 pounds. "I realized that I was much heavier than everybody else. But the wonderful thing about nursing school was that these were all people that were kind of like me. They all wanted to be nurses. They wanted to take care of people. Those values were important as a group. They were very respectful, and so I was OK there. That was really the only time in my whole life that I felt comfortable in a group, until much later."

Janet remained that weight and upon graduating nursing school she went to New York City to find work. There she met Harry. "He was Jewish and I was Catholic. He was tall and big, and he was really interesting and bright. He didn't seem to care that I was heavier, you know. At that point I still weighed about 185. So anyway, we became an item and we had a very torrid affair. I was head over heels in love for the first time in my life."

After they lived together for a while Janet needed to get out of New York. She wanted to go back to the country, where she grew up. Harry was reluctant, but decided that they should get married and then move. The time came to meet Harry's parents.

"The first time I went to visit Harry's family, he was scared to death. He said, 'My mother is going to hate you. I want you to know that.' I said, 'She's not going to hate me. People don't hate me. Don't worry about it. I'll win her over.' He said, 'She's going to hate you.' And she hated me. When I got to the door she looked at me with horror. And I had made a dress, and I had stockings and pumps on. And she said to him, "This is what you bring home to me? This is what you're going to marry?" You know, and started to faint, and left the door, and I was really upset. We went in. His father was quite nice to me, but obviously whipped. There was no way he was going to stand up to any of this. After spending the whole evening there his mom said, "If you're absolutely certain that you must marry this person, we'll pay for the wedding." And I stood up and said. 'Oh, no you won't. This is my wedding. I don't want your relatives there, because they won't like me. *You* don't like me.' "

"We got married. Harry's mother wore black. She wore a veil over her face. And when I walked out in a red dress, you should have heard her. She cried for the whole ceremony."

After they moved out of the city things started to fall apart. Harry started drinking heavily and verbally abusing Janet. They were on the verge of breaking up when Janet discovered that she was pregnant. She felt that a baby could "fix things" between them. After their son was born, Janet gained more weight. She tipped the scales at 270 pounds.

"I'm just gaining, gaining, gaining and 300 is looming. At that time everybody was doing these stomach stapling operations, and the surgeon who was doing these surgeries had been a dialysis surgeon in a hospital I was working in. I adored him. I went to see him in his office one day and I said: 'I'm out of control. I just keep eating. And I just can't do anything. And I don't want this to happen to me.' And my blood pressure was way up. So, I put it in terms of really needing to save my life. And I was still trying to save my marriage."

Janet's husband had sworn to her that "If I lost weight everything would be fine." She said, "I announced to him that I was going to do it. He was thrilled. He said, 'That's great. That's wonderful. Everything will be fine if you do this.' I felt that I would be socially acceptable. By then it had become very clear to me that society doesn't like fat people. They move away from us on buses. They move away from us in train stations. I went and had the surgery."

Janet lost one hundred pounds in nine months. I asked her how she felt. She said: "I looked like a million bucks. Everyplace I went, people were like, 'Wow, you look incredible, you look great.' On top of the world. And actually, my head was in a better space as far as my husband was concerned, because it was like, "Shape up or I'm out of here." You know, I didn't feel so desperate. 'Cause desperation was what drove me to have that surgery. I felt that nobody would love me.

"Now I weighed 170, which was still heavy, but I didn't even look like I weighed as much. So, I looked like I weighed about 140, I was normal. Wonderful. My mother-in-law was now taking me shopping. My aunt was taking me shopping. Everybody thought I was just wonderful. I had a wonderful baby. I was taking care of my husband like I was supposed to, because that's what women do. I was going to school. I was doing everything, managing it all."

I asked Janet what after effects she experienced from her surgery.

"It was painful, though that goes away. But food became a real problem for me. I couldn't eat. I would throw up. I couldn't eat half a hamburger, with no bread. And if I ate more than that, pain, pain, pain. And I would throw up. The other thing that happened to me is I got into this thing called gastric dumping syndrome, where you'd eat a little and the sensors in your stomach would say, 'Oh my God, this stomach is totally distended. She has just eaten two Thanksgiving Day meals. She needs insulin from the pancreas.' So my pancreas would dump insulin into almost no food, and I would have like an insulin reaction. I would get pale and shaky, and I would sweat and I would have horrible cramps, and the food would go just right through me. I would have diarrhea for hours. It's kind of like surgical bingeing and purging, and I hated it."

Her eating patterns changed, too.

"I couldn't eat anything I liked. Ironically, I couldn't eat apples, and I couldn't eat anything that had any bulk to it. I couldn't eat protein. I couldn't eat mashed potatoes. But what I could eat was candy, and potato chips and all this junk that makes you fat in the first place."[38]

Over time, however, her anxiety over the surgery went away. Little by little she stretched her stomach out. And Janet realized that nothing in her life was really different. "Nothing changed. When I was thin, sex was no better, nor more often, nor did my husband like me any better. His mother liked me better. I was still stressed out about my marriage and I just ate. *I gained back 100 pounds.* I was pissed, because now I was getting side effects from my stomach stapling. I was getting anemic, because when you have that much of your stomach closed off, you don't absorb any of the iron that you eat. Plus, I had this big ugly scar."

Janet left her husband. Then a terrifying medical crisis, in which her stapled stomach developed an abscess, almost killed her. Janet had to have her surgery reversed. Her story has a happy ending, however. After a long recovery, she decided to accept her body on its own terms. Her weight stabilized, and she entered a second, satisfying marriage with a man who loves her the way she is. Today Janet has a Ph.D and a successful academic career.

Janet's painful route to self-acceptance required her to reject the cultural norms she'd grown up with. She had to decide for herself what her personal shape should be. If we take a look at the everyday world of girls as they grow up and begin to develop a sense of body image, we will see how family and friends begin to echo cultural values and attitudes. As the plastic surgeon who had consulted with the mother and daughter told me, "If the daughter's friends and future husband don't recognize her qualities and if a nose job will do what it takes, she will do it. But it's mainly because of pressures imposed on her from without, coming through the mouthpiece of a mother, or any other forces."

If we examine some of these "forces" at work within the family, peer group and educational environment, we may understand how women become a "certain body."

5 Becoming a Certain Body

"My parents were always complimenting me on how I looked. It was such a big deal. I remember my father would say 'you look good, you lost weight.' And he also commented on other women, more than me. If there was a waitress, he'd say, 'Boy, is she beautiful!' Always commenting on pretty young girls. So I knew it was very important for him that I looked good too. When I'd dress up, I wanted him to see that I could be just as pretty as all those women he was commenting on."—Jane, college sophomore

As part of membership in our society, young women have to learn how "to be a body."[1] And, for the most part, what a woman observes in the mirror is what she uses as a measure of her worth as a human being.[2] We have just seen how the food, diet, and fitness industries, aided by the media, have systematically convinced women that independence means self-improvement, self-control, and the responsibility for achieving the ultra slender body ideal. But the family, school, and peer group also have a role. They reflect and frequently amplify societal norms. These social influences often take the form of rewards and punishments to urge women's bodies toward thinness. The result creates vast differences in how men and women feel about their bodies.

Seeing the Self

Growing up in American society, we are taught, of course, to value what our society values. We learn to see ourselves as others see us, in terms of social standards. Self-image develops through social interaction. According to noted social psychologist George Herbert Mead, "The self has a character which is different from that of the physiological organism, with a development all its own. The self is not even present at birth but arises later in the process of social experience and activity."[3]

Mead adds that we experience ourselves as both subjects and objects. "The individual experiences himself as such, not directly, but only indirectly, from the particular standpoints of other individual members of the same group or from the generalized standpoint of the social group as a whole to which he belongs."[4] Sociologist Charles Horton Cooley refers to this as the "looking-glass" self.[5] Our significant others, such as family and friends, are the mirrors that reflect us. What others value in us provides the basic building blocks of selfhood.

Unlike personality, tastes, and social values, our physical appearance is always visible to others. It is a critical factor in the development of self-concept for women, especially during adolescence and young adulthood.

Weight is an important aspect of appearance, affecting young women's sense of social and psychological well-being.[6] Many women experience even a few extra pounds as a major issue in their lives; they tend to weigh themselves frequently and report seeking medical help for weight problems more often than men. Although physical appearance is important for men, their traditional socialization stresses the importance of achievement (the mind) as a primary determinant of self image and self-esteem.

Women's bodily focus arises from their discussions with their friends, their interactions with family and social groups, and the messages they receive from outside this intimate circle. It is reinforced by the everyday practices that make the body central to their identity as a female—from clothing, hairstyle, and makeup; to speech, walk, and gesture. The Cult of Thinness becomes a powerful lure as society decides which is the "right" or the "wrong" body and treats women accordingly.[7]

The Good, the Bad, and the Ugly

As a culture, we associate beauty with the good and ugliness with the bad. Attractive people are "viewed as being happier, more successful, smarter, more interesting, warmer, more poised, and more sociable."[8] Research suggests that the social consequences of looking good begin as early as infancy.[9] As they enter school, less attractive youngsters are likely to be blamed and punished more often than attractive children.[10] For example, one study concluded that adults not only ascribe negative traits to unattractive children, they are also reluctant to ascribe such traits to attractive ones. Participants in this study were given written descriptions of supposed behavioral transgressions with photos of either attractive or unattractive kids attached, as judged previous by a separate group of adults. Their evaluations confirmed the researcher's hypothesis:

> An attractive child who commits a harmful act will be perceived as less likely to exhibit chronically antisocial behavior than an unattractive child, primarily when the offense is severe. Thus, adults evaluating an attractive child . . . per-

ceive him as less likely to have committed a similar transgression in the past and less likely to commit one in the future than an unattractive child.[11]

Similar attitudes apparently exist in correlating attractiveness and academic performance. In another study, 400 fifth-grade teachers examined report cards with pictures of either attractive or unattractive children. The teachers were asked to evaluate the students' IQ and academic potential. The researchers noted, "We predicted that the child's appearance would influence the teacher's evaluation of the child's intellectual potential, despite the fact that the report cards were identical in content. It did. The teachers assumed that the attractive girl or boy had a higher IQ, would go to college, and that his parents were more interested in his education."[12]

Body type has long been associated with temperament, and has given us stereotypes like the jovial Santa Claus, or Shakespeare's lean and hungry Cassius. Various pseudo-scientific attempts at inferring personality from physiology were in vogue well into the present century, including the analysis of body type. Attempts at a more scientific study of the body-personality relations date from the 1930s. Perhaps the best known example is the work of W. H. Sheldon. Sheldon adopted three categories of body type: ectomorphic (lean, angular); mesomorphic (muscular); and endomorphic (rounded or plump). They were associated—tentatively by Sheldon, but more assertively in folklore—with certain personality types. These were cerebral, active, and sensuous, respectively.

It is easy to criticize this oversimplified approach. But the general hypothesis of a correlation of body-build and personality continues to be investigated, albeit in a more sophisticated manner.[13]

Some researchers emphasize the moral implications of obesity, a legacy of our Puritan heritage.[14] The overweight body signifies immoderation , greed, and the inability to control gratification, whereas slimness epitomizes the opposite, and is attributed to strong moral fiber. (So much for the validity of the Santa and Cassius types.)

This attitude is expressed nicely by three female college freshmen:

How can people not even care how they look? It's just like they let themselves go to pot. (Not only) how they look to others and themselves, but what's going on inside of them? It's unhealthy. I'm determined to never, ever look like that.

If I was at my ideal weight I'd feel really in control of my life. Even though I'm comfortable, I still feel that if I was perfect, I would be maybe 7 pounds less than I am. And I would feel a little more positive with myself, I think. More in control.

I think being thin would just make me a better person. I just feel like I'd have more self-confidence and stuff.

Not surprisingly, weight-related aspects of appearance are more intertwined with self-concept in females than males, and they are a major factor in how others view them.[15] Being fat can stereotype people even at a very young age. One research study asked a group of youngsters to pick whom they liked best from a set of pictures of several physically handicapped children plus one obese child. The obese child was the last to be selected.[16] The stakes are even higher for females because the relationship between attractiveness and higher social status is in fact stronger for girls than boys.[17] For example, a study on the social patterns of college undergraduates found that the more good-looking the female the more she was liked by her date regardless of other factors such as personality or intelligence.[18] Women who are overweight date less often and are less satisfied with their mate.[19] In a society where beauty and charm still strongly affect a woman's social, marital, and economic success, fat women risk downward social mobility. By the same token, a man can maintain or gain status by marrying a beautiful, thin woman.[20]

My interview subjects believed in the close connection between thinness, good looks, and marriage:

> I think I have to please men if I want to get a date, if I want to be married, if I want anything, and so how I appear to men is really my final (weight) goal, like if I'm going to get married or be an old maid.

> I know it sounds corny, but if I gain weight, I won't find a husband.

Dissatisfied with Weight and Shape: A Survey of College Students

Given the premium placed on looking good, and the fear of its opposite, it isn't surprising that I found a greater degree of dissatisfaction with body weight and shape among the women in my survey of college students.[21] Their concerns with their bodies reflect large-scale trends found in other studies.[22] I discovered distinct gender differences in perception of body weight:

- Women overestimated their relative weights and thought their bodies were heavier than the medically desirable weights on the standard Metropolitan Life Insurance Co. chart. Men, on the other hand, judged their weights more accurately.

- When I asked my sample of students how much they wanted to gain or lose, the vast majority of women (95%) wanted to lose weight while the men were almost evenly split between wanting to lose and wanting to gain weight.

- Over three-quarters of the females in my sample, but less than one third of the males, ever dieted.

- Even more startling is the difference between the proportion of men and women who said they dieted "most of the time"—37% of the women and only 15% of the men.

- I asked the question: "When you look in the mirror, which best describes your feelings: proud, content, neutral, anxious, depressed, or repulsed?" I found that 50% of the men in my sample were at least content with their body image compared with only 37% of the women.

- Many women expressed anxiety about their bodies (28% of the women compared to 6% of the men). About 7% of the women felt depressed and repulsed by their bodies compared with 4% of the men.

How Much Should I Weigh?

In my interviews, it became apparent that there is no single definition of the body ideal that fits every woman and not everyone pursues the ideal with the same intensity (some women expressed their ambivalence toward this ideal, or even found ways to rebel against the Cult of Thinness). But it also became clear to me that, in general, women say they want to be thin. As Jane put it, "I'll never be happy with my weight. I will never be satisfied. I will always desire to be thin." Or Dana, "If you are thin and firm you're more socially accepted, you have more self-confidence and you can achieve your goals more easily."

What do they mean by "thin"? Most women have a very definite weight in mind. Where do they get this number? While pondering this question, I decided to go to a local diet center on initiation night. When I entered and filled out a registration form, I immediately got weighed. My weight counselor consulted a weight chart and told me that I needed to lose ten pounds. I am 5 feet 6 inches, and at that time I weighed 130 pounds. When I asked her how she decided on this amount, and where her chart came from, she said she didn't know. I asked if I could see the chart, perhaps take a copy, but she said she would have to ask for permission.

Charting Weights

The weight charts that appear in many women's magazines and elsewhere are important definers of desirable or "ideal" weight. The standard weight chart is another mirror that promotes weight obsession.

In order to measure your weight against an "ideal" weight, all you need to do is find your height and frame (small, medium, and large) and you will get a range of suggested ideal weights for your height and build. I surveyed weight charts from all the leading diet centers in the United States. I chose a weight chart taken from one of the largest diet and weight-loss organizations that is representative of commercial diet charts in the United States.

Then I compared this chart to the one used by my physician when I visited her for my annual check-up. The doctor used the Metropolitan Life Insurance Company's 1983 height and weight chart for men and women. They are the weights for a given frame and height for which mortality rates are lowest.[23] These numbers are based on medical-actuarial studies of insured men and women and therefore reflect desirability from a medical standpoint. According to the medical chart, and my doctor, my weight was within the "normal" range for my height.

I compared the two sets of charts for men and for women. I termed the diet center chart the "cultural" model of ideal weight and called the Metropolitan Life Insurance's chart the "medical" model of ideal weight. These two charts appear in Figure 8 (for men) and Figure 9 (for women). If we compare these two figures, we will see that they reveal a large gap between the two charts for women.

On the average, the difference between the two charts for men is about 5 pounds, whereas the difference is close to 20 pounds for women.[24] This difference represents, in pounds, our culture's pressures on women to be thin.

I wondered if women really *wanted* to be at a "culturally desirable" weight. So I asked the students in my sample: "How much weight would you like to lose or gain at this time?" Each response was added to, or subtracted from, the student's actual reported weight in order to determine desired weight. I plotted the students' desired weights and actual weights and compared them to the cultural and medical chart (see Figures 8 and 9).

The graph clearly illustrates how women's average desired weight gravitates toward the cultural model of ideal weight, rather than toward the medical model. For men, the deviation between actual and desired weight is so minimal that comparisons are difficult to make. On the average, there was approximately a one pound difference for men between their actual and desired mean weight compared to a ten pound difference for women.

For some height categories the men's mean desired weight was *heavier* than their mean actual weight.. This isn't surprising. Men are taught that being big is one way of being powerful, and they often confuse weight with build and may avoid dieting because they believe that it will reduce their strength and virility.

In the meantime the average American female has become heavier.[25] A woman of average height and build will find it difficult, if not impossible, to meet these stringent and increasingly elusive cultural weight norms. As we

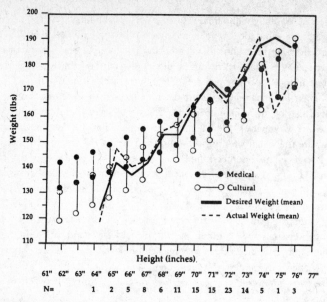

Figure 8 Range of actual and desired body weight of male college sample compared to cultural and medical standards

will observe in the following memoir, our culture sets up rewards and punishments to ensure women's lifelong involvement in becoming a certain body.

Growing Up Fat: The Story of Rene

When I interviewed Rene, she was a graduate student in her late thirties. An attractive woman with a great sense of humor, she is 5 foot 6 inches and weighs approximately 160 pounds. She was eager to tell her story, hoping that younger women might benefit from her experience.

Rene is a first-generation American. Her young parents fled Europe during World War II, and she was born in the United States. I asked Rene about her first memories concerning her body image. "Early pictures of myself show I was a normal-sized 4 and 5 year old," she said. "I had big bones, I'm dense, you know, I'm endomorphic, but I was not fat. But when I was growing up I got very mixed messages from my mother and my grandmother. They would say, 'You need to thin out, you know. Hopefully, when you grow taller, you'll be a little thinner.' And then of course, my mother and grandmother would play it off on one another, doing their own mother-daughter thing. If my mother was on my side, my grandmother would say I needed to go on a

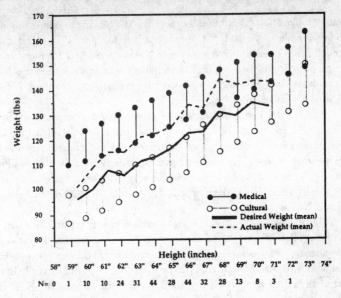

Figure 9 Range of actual and desired body weight of female college sample compared to cultural and medical standards

diet. If my grandmother was on my side, my mother would say that I needed to diet. But no support ever."

Rene remembered the first time she felt bad about her weight. "I was in the sixth grade, and you know how girls are in the sixth grade. It's just a hard time, and everybody is very cliquey, and I never could figure that out. That social stuff was very difficult for me. I remember walking past some girls on my way to the bathroom, and they started singing the theme song to the TV show *Rawhide*. I can hear it in my mind—it was about cowboys herding cows—'Rollin', rollin' rollin', keep those dogies rollin', keep those dogies rollin', rawhide' . . . I pretended I didn't hear it, put my head up, and just walked right past them into the bathroom and walked right past, back again. I remember feeling horrible about that."

She also recalled a dance she went to her freshman year in high school. "My grandmother came to my friend's house to see us go to the dance, and the two boys came there to pick us up, and we were dressed early and we came down. My friend Jill was blond and tiny, and she had on a white dress with roses on it. And I had on a royal blue satin dress, that of course you had to buy in the women's department, so it was a much older-looking dress. My grandmother had difficulties with English and she said, 'Oh, Jill, you look like a flower' in her heavy accent. And looking for something nice to say about me, she said, 'Oh Rene, you look like a tree,' thinking it as a compliment. I remember the scene to this minute. And I remember my friend not knowing how to react. And I didn't react at all. I mean she didn't know what she was doing, and I was busy mediating, and saying 'It's OK Grandma, that's fine.' "

Rene paid a high price for her perceived unattractiveness, not just in terms of her emotional and social well-being, but in her educational life too. The unconscious favoritism shown to attractive school children, mentioned above, may ensure their higher degree of academic success. Rene fulfilled those expectations. Ambivalent about drawing attention to herself, when she got to high school her grades dropped. She was told, and she felt, that she was "dumb."

"My grades were poor and my two friends were very bright. They were in National Honor Society. And the teachers kept saying to me, 'If you would pay attention, if you would concentrate, if you would do your homework, if you would try harder, you could have straight A's.' And I never believed them. Because it was too scary to think about being outstanding. But I also kept getting mixed messages. The guidance counselor told me when I was a junior that I could not go to college because I was not smart enough. The same year I was a National Merit Scholar because I scored so high on the test. Here I am with a 71 average in high school, and now I'm top 3% of the country. But then they said the test was a fluke. 'You test well , but you're not really smart.' That's what I was told and I believed it."

In the meantime she struggled with her growing anger at how she was treated, and she looked for acceptable ways to develop herself and channel her feelings.

"I was always fighting. I was always kind of waiting for somebody to attack me, and just wanting to come back at them. So I think my adrenalin level has kind of chronically been up there."

But at the same time she tried to compensate for her bad feelings about being overweight by being "everybody's buddy." "I was also the mediator. It's always been my role. And I never had dates. Never had any dates in high school, but I hung out with the guys, you know, I was always their friend. I knew that I was never going to be a cheerleader, and I was never going to be anybody that when I walked down the hall, they were going to say, 'Wow, isn't she beautiful.' What they were going to say is, 'Gee, she's a really nice person. I really like her.' And that became what was important to me, that I was a nice person, and that I was a good person."

In many ways Rene did not take well to what she perceived as a traditionally feminine role. "In terms of all the games and the complexities of being a girl, I never bought into that. It didn't make any sense to me." She remembered wanting to be a boy for a long time." When I was in 5th and 6th grade, I played Little League baseball because there was a new coach in town, and nobody told him that I was a girl. The boys all knew that I was a girl, but I cut my hair really short and I wore clothes so that I didn't look like a girl. And I was the star catcher."

"Boys weren't terribly complex. You knew what they were thinking because they were pretty much head to mouth. There wasn't a lot of game-playing, you know. You knew where you stood. And also, since I wasn't date material, they didn't care if I was fat. I was OK."

She described how she tried to negotiate a gender role identity for herself that was a mixture of what was culturally expected of boys and girls.

"The male side of me became very developed. Because also my bravado fit in with being a boy. You know, like 'If you don't like the way I look, then don't look.' That was very boy-like. But I was very nurturing, very caring, and very sensitive, which are girl characteristics."

Eventually, Rene was able to develop a self-identity and self-esteem outside of society's expectations. She ignored her guidance counselor's advice, went to college, and discovered that she could excel academically. (She has completed her doctorate in the health field.) She found that she had her own kind of femininity and attractiveness, and is now happily married.

The emotional pain of her journey shows the importance of physical appearance, especially weight, in the development of female self-concept. Even for average-looking women, it is difficult to grow up in a culture of such high standards and expectations. For those who are overweight, it is an especially punishing experience.

Thin Promises: The Rewards of the Right Body

Women continue to follow the standards of the ideal thin body because of how they are rewarded by being in the right body. Thinness gives women access to a number of important resources: feelings of power, self-confidence, even femininity; male attention or protection; and the social and economic benefits that can follow. The students I interviewed knew exactly what thinness meant to them. Julia commented on how becoming thinner really made a difference in getting men to respond to her.

Last summer when I lost a lot of weight, men were much more receptive to me, and it's flattering. I kind of like the feeling how a woman can tease men almost, you know, just a look, or what you're wearing, or the way you look at them, and that's exciting. I mean, it could be dangerous, and you don't want to push things too far, but again it's exciting. Because I've never experienced that before, you know?

And Elizabeth said, "When I lose weight, I have a wonderful feeling of power. It's like I am in control of my body . I'm thin, and it's so great. I even commented on it one time to this friend of mine. I said 'Charlie, it's so funny, you lose 30 pounds and guys really want to pay a lot of attention to you,' and Charlie said, 'Yeah, it's kind of lame' And it *is* kind of lame, but it doesn't surprise me at all. Losing weight was great for my ego."

For so many women whose bodies are their primary identities, the Cult of Thinness promises the rewards of cultural acceptance.

6 Joining the Cult of Thinness

"I was raised by a domineering mother and also a strong father who didn't give me very much personal space to develop, so I was really used to being told what to do. I learned from a very young age to surrender myself to other people's will, desire, and wants . . . to really set myself aside."—Anna, former religious cult member

"I see commercials with these bodies and I want to look like that. I have this collage in my room of just beautiful bodies, beautiful women. And at the bottom it says 'THIN PROMISES' in really big letters. I have it up on my mirror, so I look at it every morning, just to pump me up a little bit, motivate me, dedicate me."—Elena, college sophomore

For Elena, the pursuit of thinness has taken on the qualities of a religion. She makes a collage out of media images, paragons of female beauty. These are the "totems" that she worships; the inspiration for her quest for an ideal body. Her mirror can be compared to an altar, where she examines herself and fervently prays that she will be able to attain her ideal through practicing the rituals of dieting and exercising. Her daily mantra, "Thin Promises," keeps her dedicated and focused on a physical self that must be continually improved.

The college-age women I interviewed related their struggles and rewards and disappointments as they learned the culturally accepted ways of "being a body." They told me about the body watching and food monitoring practices they followed. These ritual aspects serve as powerful anchors to membership in the Cult of Thinness.

One of the most common practices is body monitoring—carefully scrutinizing the mirror for one's own physical flaws, or examining "the competition" for comparisons and defects. This involves treating the body as an object, in fact, experiencing oneself as "unembodied."

Body measuring techniques and incentives, along with food watching rituals, keep women like Elena "pumped up" and dedicated to the cause. In some cases these rituals can be painful reminders that one is not meeting these standards, a failure akin to eternal damnation. They require a great deal of time and energy, even a reorganization of daily life, not unlike the practices that religious cultists like Anna followed. "We would get up at 4 a.m. to meditate and do yoga for 2½ hours every morning," she told me. "It was hard, and everybody watched to see who showed up, who stayed awake, who practiced correctly. People felt guilty and did self penance if they didn't follow the practice."

Many Ways to Measure Up: Mirrors, Clothing, Photographs, and Scales

Methods of evaluating the body can be precisely quantitative—the reading on the scale, the image in the mirror, or the inch in the waistband of one's skirt. There are also more subtle indications, such as the admiring or critical glances and remarks from friends, relatives, or boyfriends.

Body comparisons are also ways of measuring. Very often such comparisons serve only to increase a sense of competition and insecurity in women. Do they measure up to other women in their immediate circle of friends and relations? Then there are the more global measures of comparison that are present in the wider society. Women may judge themselves according to the perfect images used in advertising or the media, or compare themselves to high profile women in beautiful-body professions, such as acting, modeling, ballet, or gymnastics.

Many of the women I interviewed looked into the mirror and believed they did not measure up to the societal expectations of the correct body image. "Not measuring up" sometimes led to strong feelings of self-hatred.

Cathleen reacted to these feelings by going on an extreme diet that led to anorexia: "I will never be satisfied with what's in the mirror. When I see other women I want to be better, thinner, than them. I would rather be anorexic than not."

Other women look in the mirror and enter the purgatory of self blame, for not being able to control their appetites. Lisa said: "I think the real problem is my whole self image—the way I see myself—never being able to achieve the goal of looking like those women body builders or being able to control myself. I feel so weak—'there you go again giving into your eating problem.' I'm just so powerless. I feel awful about myself, pretty much hate myself if I don't look a certain way. When I'm home and I'm gaining all that weight, I feel like shit. I avoid mirrors."

Roberta told me: "One day this week I woke up in the morning and I looked in the mirror, and I just was disgusting looking. And I just went right back to bed. I missed class, I could not get up. I can look in the mirror and just go, 'Yikes!' I guess I really feel like if anyone's going to like me I'm going to have to be beautiful."

Many of the women I talked with used the fit of their clothes as a way to watch their bodies, as well as an incentive to maintain or to improve their shape. When the clothing was loose, it was a time for celebration; when the fit felt tight, it was a source of emotional pain. Getting into the right size can be an aid in losing weight for, as one woman put it, "You've got to find yourself some incentive clothes." Many expressed certain measurements they would tolerate in clothing and swore to stay within these sizes.

"You know there are limits," said Angela. "I don't want to go above size 10 or whatever—I'd like to stay at my size 8."

Jane said, "When I notice that my pants fit a little tighter or something like that, then I'll stop and go to my conventional diet where I just don't eat as much as I have been."

Judy noted that when her skirt was tight or her pants no longer fit, "I snap at everyone and I'm cranky and miserable. And when my clothes fit right and I walk down the street and I know my ass wiggles—I feel great. Everything's right. It's like a high."

Another important yardstick for body measuring comes from photographs taken over a period of time. Women examine these pictures in excruciating detail and evaluate their body shape over the months and years. Several students I interviewed frequently compared their pictures from high school with their current college pictures.

For instance, when Tina, a freshman, returned home for Thanksgiving, her father "just looked at me and said, 'I can't believe you're this heavy.' And my mother too. I looked like a totally different person. Within a matter of one semester my face was plump. Lately I've been looking at my high school pictures. I had my best year senior year—I was the best body. Now I just look at those pictures and I think, 'I could get down to that weight if I really wanted to.'"

"When I picture the body that I want," Rita said, "I picture myself during my freshman year in high school. I have a picture at home. It was of me standing there in a move from a cheerleading routine. You could see the bones. I wasn't anorexic looking, I was just tight. I was thin, but everything was in place. Tight and muscular."

One of the most obvious body measuring devices is the scale. In a sense it is the totem of all totems, encountered at gyms, fitness clubs, and diet centers. It is often a household shrine as well, with its own prescribed daily ritual. Most women disrobe before going on the scale, stripping themselves of any excess weight like jewelry or hair clips in hopes that they will "measure up" to certain weight expectations.

Joan noted: "The big thing at my house is getting on the scale every day. My mother will ask 'What do you weigh?' and it's a big thing. We weigh ourselves separately, but the question always comes up during the day."

Ruth told me: "I get on the scale five times a day. I just get really nervous. I panic when I get above 120. A lot of times, like last night, I'll go out and exercise right then."

For Marina, seeing extra pounds on the scale was like a death warrant: "I gained weight and I didn't feel good about myself. I gained four pounds. Doesn't that sound stupid? I say that to myself, but when you see it on a scale, it's like death. It's like someone scraping their nails on a chalkboard. It's like you can't go up there."

Measuring Up to the "Pros"

Certain body-conscious occupations create other sub-cults, whose participants find that they must continually work at the "right" body. For most of Cindy's life, wanting to be a gymnast meant constant attention to body work. A college junior, she told me she had been extra conscious of her body image in high school because this sport demanded a certain height and weight limit.

"In gymnastics, they made us so weight conscious. Every single day we weighed in. Once, the day before an important competition, one of my friends weighed in over five pounds. The coach told her that if she didn't lose the weight by the next morning, she couldn't compete. She did everything she could. She put on a [sweat]suit and ran all night. She had beautiful long hair but she just chopped it all off, and the coach didn't care—as long as she lost the weight. Whenever there was a weigh-in, if you walked into the bathroom there was someone puking or on Ex-Lax."

Cindy started gymnastics when she was five. She practiced for four hours a day, six days a week, and on most weekends she was away at competitions. Ultimately, such an obsessive focus destroyed her pleasure and interest in the sport, but not before it had also stunted her physical growth.

"My father is over six feet. My mother is 5 foot 9 inches and my sister is also tall. I was supposed to be tall too—I wear a size eight and a half shoe. But my doctor said the hard training stunted my growth. I am only 5 foot 3 inches."

During her long years as a gymnast Cindy ate what she wanted. Her mother never objected to this since she was getting exercise. "Typically, I would get home from school and have dinner about 3:30 P.M. by myself. Then I'd go to the gym. So I never really ate with my family. On weekends we tried to get together on Sundays and sit down and eat. If I hadn't been in that kind of active sport, burning everything up so fast, I think my body would have been like a normal person. I never thought about food when I was younger, because it was never something that I wasn't allowed to have. I was the good girl and I stayed in gymnastics for two years longer than I wanted because my parents wanted me to. If it had been up to me I would have quit sooner, a lot sooner. I started to burn out. My freshman year of high school was my peak."

When Cindy got a bit older, around sophomore year, her coach started to take food away from her. "My coach was saying 'You have to lose weight before you compete, or you can't work out.' He was taking it away and when my mother found out about that she would just oversee it herself. She would say, 'You can't have this, you can't have that.' I'd find a way to binge. I'd go into my room with a bag of cookies and I'd shove them all in my mouth. She never knew. This was a big problem for me because I am a binger. I'll binge and then I'll starve myself the next day. I've been through it all. I've been through the bulimic stage. I've been through the Ex-Lax and I've been through bingeing and starving. When I was really thin, my junior year of high school, I looked good. I got compliments from everyone." Cindy's years of rigid discipline and practice backfired into an eating disorder, and has continued to be an issue for her in college.

Family Input and Peer Pressure

Early on, parents, siblings, peers, and even family doctors and diet clubs are important "guides and gurus" in the process of body watching. The women I interviewed told me that what their parents thought of them had quite an impact on how they perceived themselves. Helene felt her mother was always monitoring her:

> My mother wanted me to have everything she never had, like a college education. She wanted me to have more than she did. But she was critical of my body. If she didn't like what I was wearing she'd tell me right off the bat, 'Don't you know how to dress?' Last year she called me a moose, and that hurt. Sometimes I think I need that, like just to make me aware so that I do something about my weight. I can see that all her pushing me has gotten me where I am. At my house we get on the scale every day. And my mother will ask, 'What do you weigh?'
>
> When I'm home I drop weight, because my mother is always on my back. When we go out to eat she tells me what I should order. When I look fine, my mother says nothing about my body, not even a compliment. But when I start gaining weight, the criticism begins.

Peggy's mother took her to the doctor when she was not getting rid of her baby fat. Her mother put a high premium on looking good, and expected her daughter to do the same:

> You know, my mother thought I was fat. I was 11 years old at the time. She took me to her weight doctor. She put me on extensive diets and I didn't like that. She gave me these amino tablets and she'd search around and make sure I didn't take

cookies with me to school. She watched me. I actually remember the moment when I first felt shame. I was undressing, and all of a sudden, I jumped away from the window, and suddenly realized I should cover myself up.

Andrea's mother was also concerned about her daughter's weight. Andrea admits that she was, and still is, pudgy. Her baby pictures show her as a chubby little kid. She said:

> When I was smaller it was overlooked. You know, you're cute and dimply. But then as I got into grammar school I became heavier. From that point my mother was always concerned that I was heavy. She'd say, 'Couldn't you go on a diet?' And she tried her best to be nice about it, but when you're heavy you don't care to hear it from anybody. I joined a diet club when I was in high school because my mother said I had to. I was practically physically dragged to this place, totally against my will. I was so embarrassed. But I looked at everyone else who was there, and thought, "Well at least someone's bigger than me."

Many women felt that peer group interaction was an important indicator of how their bodies measured up. June relates her discomfort during the transition from elementary to middle school:

> Because I was taller than all the guys in middle school, this one kid used to call me 'Amazon.' When I look back on it, the reason he called me that was because of my height, but I thought it was because I was fat. I really was never overweight as a kid, just the tallest. But then, when I went into 7th grade, I had this incredible will power. I just didn't eat.

Dangerous Comparison and Deadly Competition

Snow White's stepmother, the Queen in the Grimms' fairy tale, prided herself on being the fairest in the land. She consulted her magic talking mirror for reassurance every day. But when Snow White grew up to be more beautiful, as confirmed by the mirror, the wicked Queen ordered her killed. In the end, of course, Snow White was rescued by the King's son, and the cruel, jealous Queen met a grisly end.

Like the stepmother, many women perceive others as more attractive than themselves and feel envy, rage, and even violence toward one another. Good-looking women, regardless of their other attributes, are just more competition for the few Princes out there. My interview subjects reported that they constantly compared themselves to their sisters, mothers, and girlfriends. When they felt they didn't measure up to the competition, their anger, resentment, and even evil thoughts rivaled those of the wicked Queen.

Cory felt her body did not conform to the standards of beauty set by her culture, and was angry and despairing that the mirror did not tell her what she wanted to hear.

> I think I am not attractive.. I hate my skin and I hate my body. I'm small-chested, and have big hips and cellulite on the back of my thighs. It's disgusting. I see girls in my classes and I think they are very pretty. They've got gorgeous long hair, blue eyes. I tried to change myself in addition to my weight. I've tried to comb my hair long, and it doesn't look good. I've changed the color of my hair. Nothing is right. I don't like my nose, I don't like my face and I hate my double chin.

Beauty pageants are a good example of how society fuels the fires of competition and envy. While giving a passing nod to talent contests and oral interviews, pageants still focus on the ritual line-up of bodies in swimsuits and gowns. One student related her experience:

> Last year I won the local competition and I went on to compete in the state competition. My agent said I had a chance. As soon as I got to the contest, I realized that girls just enter year after year until they win. And I was one of the few people who was there for the first time. The girls were so bitchy and so catty and not friendly at all. A big part of winning is showmanship. It's how you approach the judges.

Some women described their transition to college as also involving a heightened need to compete with yet another set of beauty expectations.

Molly said, "I feel the competition here to look good. I mean you've got some incredibly good looking women that go to this school. They call it 'the beautiful school.' I never felt like I had to compete until I came here and people put all this pressure on you."

Sometimes the competition for male attention brings confrontation and the threat of violence. Mary was at a bar with her roommate, and narrowly avoided a fight.

> I was at a bar hanging out, and there were a couple of guys on the other side. My roommate called me over to introduce me to a couple of her friends. I didn't look good, I had a bandanna in my hair. My makeup was all over the place. I mean, I'd been through a lot that night. So I was talking to these guys, and this girl comes over and nudges me. She was made-up, her hair beautiful. She said, 'Excuse me, but are you with this party? Cause if you're not, would you mind moving?' That was it. I completely lost it. The next thing I knew, I was in this girl's face and said 'Who the hell do you think you are?' And my roommate just pulled me out of the bar because there was going to be trouble. This girl thought she was so much better than me. The guys just stood there. They were all like, 'You don't want to mess with her, you know?"

Sometimes the competition hits close to home, when mothers, daughters, siblings, and peers are comparing their bodies. "My mother was a model," Lucy told me. "She was always Little Miss Beauty Queen. And there was pressure for me to follow in her footsteps. But my mother was jealous of me. I could feel that tension between us. She would often say, 'Well, you look a little scruffy today.' "

Irene was competitive with both her mother and her sister:

When I was in high school my mom dyed her hair blonde, and was skinnier than I, and I was incredibly jealous. It drove me crazy. I didn't want guys I liked to meet her. I didn't really tell her, but on the beach she'd wear bikinis, and I hated it. I thought my mother was trying to be a teenager when she was older, and I was really jealous of her. I like being with my father alone rather than with my mother, because I guess I feel that I'm still competing with her.

What got Irene particularly upset was that her mother could eat almost anything and never put on weight. Irene's sister was also thin and had her mother's metabolism. Irene was built more like her father, who was big-boned, tall, and somewhat overweight.

"My father used to say that I had no style. My sister had all the style. She could wear anything because she was so thin. I always got A's in school, and I was in Honor Society, whereas my sister did badly. But that was the least of my concern, getting A's. I wanted to be thin. I would just look at my sister, and think, 'My god, if I could only be thin like her.' "

Peer competition intensifies in a closed community like a boarding school. Maria talked at length about her experiences at the school she attended in France before coming to college. The girls in this boarding school practiced body and food watching rituals with great concentration, as befitted their exclusive cult subculture.

The first year there, when I was fifteen, I didn't have any problem with my eating. The second year, Sandra came along. Sandra is German and very pretty. She and I became good friends. We had both been friends with Claire, who had steadily become anorexic at school, then died that summer. It really didn't hit Sandra and me. We had gone on a diet the previous year and continued this at school the next year. First of all we started purging (vomiting). We'd actually gotten this idea from Claire. And it got to the point where it would come naturally. I was bulimic for six months. It was horrible. My housemaster noticed, and sent me to the doctor. I had a lot of chest pains at the time—I had just destroyed the lining of my stomach. They gave me this awful white liquid. And I wasn't necessarily thin—I was pretty much the same weight. Then the following year, as seniors, Sandra and I started with this vicious competition about losing weight. I ate only carrots and this low fat cream with diet crackers. I started losing weight and Sandra was losing weight faster, because she was thinner than I was to begin with, and taller. So everyone was yelling at me, 'Why are you letting Sandra do this to you!"

At this boarding school, appearance is everything. Everyone there was gorgeous. People look at my yearbook and say 'wow!' And I think that's the way the school wanted us to be—glamorous and gorgeous and intelligent. The perfect girl at this boarding school would be someone with long blonde hair, terribly skinny, big blue eyes, some sort of affected accent, and no makeup. Everyone tried to put makeup on to look like there was nothing there. They would color their cheeks, or whatever. Girls wore colored contact lenses at the age of 13. Everyone also definitely tried to be skinny. The girl next door to me was anorexic. She used diuretics. Another friend upstairs was also anorexic and she had to go to the hospital because she had used laxatives for so long her body would no longer function normally. Last year when I went back for a reunion I found out that another good friend of mine, Carla, had died of an eating disorder.

At this school, if you had dinner at the table and ate the entire meal, you were considered a pig, and you were going to be talked about. The biggest thing was to just run around with the salad bowl and ask if any one had any extra salad.

We lived in a girls' house but there were guys there as well. That's who you had to look good for. It doesn't hit you how bizarre something is until someone from the outside says something. We lived in such a closed community. You weren't allowed to go into the city, you weren't even allowed to go to any of the towns. You had to stay there.

Measured by Men

One all-important indication that a woman has the culturally correct body image is the attention she gets from men. This may take the form of getting a date or merely prompting a nice remark from a male concerning her appearance.

Virginia was sure that her weight problem was responsible for her empty social life: "Now, looking back at pictures, I looked fine, I was never even overweight. But at that time I thought, 'Oh I'm horrible and that must be the reason I don't have a boyfriend.' "

Judy felt pleased to be recognized as attractive by one of the waiters at a restaurant she frequented with her college friends: "The host called me over last night, and he said, 'You know the waiter thinks you're very attractive. Is there a chance you'd say yes if he asked you out?' So if he calls I will go out with him. It really made me feel good last night because I was out with five other very pretty girls, and he picked me, thought I was attractive. It's usually my roommate that gets the guys."

Food Watching Rituals

The ritual of food watching goes hand in hand with body watching. It is often practiced on a daily basis, either alone or with the help of another individual (usually a family member) or with a group such as a diet club. Food watching ranges from calorie counting, to full scale dieting, to the behavioral symptoms of anorexia or bulimia.

Almost all women in my sample of college students had been on what they termed "a diet" during their teenage years. They consider dieting normal behavior.

For example: "I would do just normal dieting and then exercising, nothing crazy, but just cutting down, maybe more than I should actually. I'd just eat very little from each food group, and exercise a lot. I was always hungry."

And: "I never went to Weight Watchers or anything. I went on the Sherry Vitti diet a couple of times. You know, no junk, nothing between meals, don't eat after supper, that kind of thing. I would eat three times a day but small amounts. And sometimes like at night, I remember the room would start to spin a little bit, and then I'd have to get some whole milk or something to put in my stomach."

Some dieting behavior consists of fasting for days at a time. Georgia went on a fast after her boyfriend made a comment concerning her weight. "I was so hurt at the time, so mad at him. What I did was just not eat anything. I went to aerobics every night. And of course there was nothing else in my life. I had my school work, but I had no other interests. I'd go out on weekends with my friends, but I didn't even want to do that much because that was always sitting around food. It was great while I looked good and I got all the compliments, but I don't think it was worth all the anguish and deprivation I went through. And I hadn't changed my eating problems, I had just suppressed them for 8 weeks. I lost 30 pounds. I remember being exhausted when I got home from school. I wouldn't feel like doing anything. Once I started eating, I gained the weight back again."

Parents, friends, and diet organizations can also be involved in "helping" women food watch. Judy said:

> My mother was very critical of my appearance. I was always the fat one and she was the thin one. She would say that I have the fattest thighs in the world, and that I'd better watch what I'm eating. She would always make sarcastic remarks. She would say, 'If you want to diet, I'll help you. I'll make special meals for you. I'll do anything I can for you.' She was good in that aspect, but in the back of my mind I knew she was always going to say something when I picked up that Twinkie.

Good Food, Bad Food, and Disciplined Eating

The art of food watching consists of taking on rigid attitudes toward food, in which all food falls into "good" and "bad" categories. The good foods are those that are nutritionally sound. Bad foods are the ones that tempt you to overeat, that can make you gain weight. They include sweets and anything with the "F word"—"fat.""

One student mentioned how she would have to sneak these "forbidden" foods during the night in her household. Her mother would get upset if she found her eating sweets. "I would wait until everyone was in the bedroom watching TV or something and then I'd go into the kitchen, maybe steal a piece of pie. If we were having a main meal, you could have a second helping of meat or some more vegetables, but don't go for that second piece of pie. She would make mine really small. My mother was always watching me."

Emily contrasted a good eating day with a bad one. "When I'm trying to eat well, I'll have my carbohydrate in the morning. And for lunch I'll probably have a salad. Then for dinner I'll have a nice piece of fish with vegetables and a salad. When I want to eat well, I can. But I have bad eating days. That's when I have a bagel with eggs and a banana. For lunch I'll eat something fried, or a sandwich. Then have a big bag of chips and then will come dinner. A potato and meat with a huge salad with everything on it including dressing. And then going to the sweets and treats and getting six chocolate chip cookies and going home and having late night snacks, like nachos with cheese or another huge sandwich. That's bad eating."

Restricting food intake requires that women eat in a highly disciplined way. Counting calories is an important ritual in maintaining this discipline; making sure not to transgress the limit; following the scriptures of the diet bible. The women in my sample who went off their diets by eating forbidden foods felt as if they had broken the rules, or had fallen from grace. They had to confess the sin and make plans to atone.

Hillary, for example, felt guilty if she ate food considered fattening, like cheese. "If I put a piece of cheese in my mouth I feel awful. I feel fat, I mean enormous. I've totally messed up. I'm ruining my body. I'm going to be two pounds heavier tomorrow because I ate this cheese. And this is going through my mind, and it's becoming this big drama. It's really weird."

Hillary would do penance for this sin by not eating anything the next day. The extremes of this kind of eating verge on the dis-orderly, as we will see in the next chapter.

7 From Disorderly Eating to Eating Disorder: The Cultural Context of Anorexia and Bulimia

It's Your Duty to Be Beautiful

> Keep young and beautiful.
> It's your duty to be beautiful.
> Keep young and beautiful,
> If you want to be loved.
>
> Don't fail to do your stuff
> With a little powder and a puff.
> Keep young and beautiful,
> If you want to be loved.
>
> If you're wise, exercise all the fat off.
> Take it off over here, over there.
> When you're seen anywhere with your hat off,
> Wear a marcelled wave in your hair.
>
> Take care of all those charms,
> And you'll always be in someone's arms.
> Keep young and beautiful,
> If you want to be loved.[1]

The words of a popular song echo a powerful message in our culture— that only the beautiful and the thin are valued and loved. Reiterated by families, peers, and the school environment, this notion is taken seriously by many young women. So seriously, in fact, that anorexia nervosa (obsession with food, starvation dieting, severe weight loss) and bulimia (compulsive binge eating, followed by purging through self-induced vomiting or laxatives) occur ten times more frequently in women than in men.[2] These syndromes usually develop during adolescence[3] and, until recently, were more prevalent among upper- and upper-middle class women.[4]

These behaviors carry long-term physical risks, ranging from tooth enamel destroyed by stomach acid, to malnutrition, to organ damage, to death. Anorexia is one of the few psychiatric disorders with a significant mortality rate. The American Anorexia/Bulimia Association estimates that

10% of those diagnosed with anorexia may die.[5] Bulimia is thought to be four to five times more common than anorexia, but is more difficult to detect. Bulimics are usually secretive about their gorge-and-purge episodes, and since there is often nothing about their external appearance to alert anyone to the presence of the disorder, the condition goes undiagnosed unless they seek help themselves. The number of women dying from bulimia is hard to estimate, but bulima can have serious medical consequences, like gastrointestinal damage. The emotional toll of these disorders can include feelings of despair, self-loathing, guilt, depression, low self- esteem, and an inability to conduct normal relationships.

In recent years, the topic of eating disorders has emerged from the realm of clinical case studies in scholarly journals to a place of prominence in the public eye. As is often the case, the afflictions of celebrities have helped pave the way for a greater focus on a widespread problem. The starvation-related death of singer Karen Carpenter, and the confessions of Jane Fonda and Princess Diana that they suffered from bulimia, have received copious coverage and created a climate of acceptance.

The Cultural Link

Earlier, in Chapter 5, I described the differences between the cultural and the medical models of ideal weight. When I surveyed the college women in my sample, I wasn't surprised that 77% of them chose the cultural model as their desired image, and 23% the medical model. But I was interested in the link between the culture's demands for slenderness and the rise in eating disorders. I wanted to know if adhering to these norms might make these young women more vulnerable to developing eating problems. So I administered the Eating Attitudes Test,[6] a standard measure of eating disorders. The results gave me the connection I was looking for. Of the women who followed the cultural ideal, 24% *scored in the abnormal range*, compared to only 8% of the medical model followers. In addition, almost half of the cultural ideal followers (47%) compared to only about one-fourth of the medical followers (26%) reported significant to extreme concern about their body weight; 34% said they felt anxious, depressed, or repulsed by their bodies, compared with only 25% of medical model followers. The link was clear: *those women who believed in the cultural definition of body image were more at risk for the development of eating difficulties.*

If we were to rely only on traditional psychology to explain the current near-epidemic increase in eating disorders among college women, we would have to assume an increase in the underlying mental and emotional features that produce such symptoms.[7] There is a big difference, however, between disorderly eating patterns and clinically defined eating disorders. College students in my sample displayed many of the *behavioral* symptoms associated

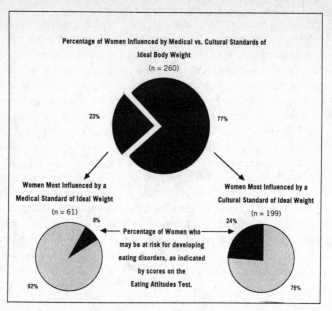

Percentage of Women Influenced by Medical vs. Cultural Standards of
Ideal Body Weight

(n = 260)

23%

77%

Women Most Influenced by a
Medical Standard of Ideal Weight

(n = 61)

8%

92%

Percentage of Women who
may be at risk for developing
eating disorders, as indicated
by scores on the
Eating Attitudes Test.

Women Most Influenced by a
Cultural Standard of Ideal Weight

(n = 199)

24%

76%

Figure 10 Percentage of women influenced by medical vs. cultural standards of ideal body weight who may be at risk for developing eating disorders

with anorexia and bulimia. In order to live up to the cultural mandate of thinness, they engaged in calorie restriction, chronic dieting, bingeing and purging, and the use of diuretics or laxatives. Some also used extreme exercise to control their weight, becoming overly dependent on a rigid workout schedule to make them feel "alive." Yet, they did *not* exhibit the full constellation of psychological traits usually associated with an eating disorder, such as maturity fears, interpersonal distrust, and perfectionism. Their behavior mimicked anorexia and bulimia without the accompanying psychological profiles.[8] Some researchers refer to this pattern as "imitative anorexia," "subclinical eating disorders," or "weight preoccupation."[9] I refer to it as *culturally induced* eating; a pattern of eating disordered symptoms in otherwise psychologically "normal" women. Disorderly eating and obsession with food is a widely accepted way to deal with weight and body image issues. *It is normative behavior for women who are part of the Cult of Thinness.*

But sometimes these strategies have unintended consequences. Severe food restriction over time may trigger an uncontrollable binge, which the dieter feels necessary to then purge. Women who tamper with their body's natural metabolism through dieting may find they gain weight on fewer calories.[10] Excessive exercise can lead to injury or burnout, or even halt menstruation. All of this fosters constant weight preoccupation and an overconcern with body image. These behaviors may lead to depression, or develop into long term eating disorders with their accompanying psychopathology.[11]

There were indeed women in my research sample who would be classified with an eating disorder. Frequently they were women who had a history of severe trauma—sexual abuse, family dysfunction, and/or physical abuse. They used eating as a mechanism to cope and to empower. In a sense, manipulating food intake is one culturally approved way that women can gain some influence over their environment. Control of their own bodies is a substitute for control over their economic, political, and social lives. For example, rejecting food may send a powerful message to a problematic family. Overeating may be a safe way to soothe emotional pain.

Linda, one of my interview subjects who was bulimic, told me:

> Food is nurturing for me. Over the past six months, since I've been dealing with sexual abuse issues, my bulimia has disappeared to the point where I can go for a month without having a problem. When I do have a problem, I know what I'm doing with the food. It's nurturance, and I understand that, at the time, and that's when I give in to it. You know, you can't fight your battles on every front all the time. I realize I'm giving in to this bulimia, but if that is what I need right now to take care of me, that's all right.

Purging may be a means for some women to vent the anger and frustration they feel in dealing with their home environment, sexual abuse, or mistreatment from society. They may experience secondary advantages such as weight loss, but this may not be their primary motivation. Some women may feel an important sense of control in the decision to eat or not eat. Food is a means for self-expression and power.[12]

Kim, an anorectic, expressed the difference between disorderly eating and eating disorders.

> I think my issue was wanting to control my life. There were a lot of family issues and personal issues that were going on in my life freshman year of high school and I just started with a diet. I suddenly decided I wouldn't eat more than 300 calories. My parents had gotten separated freshman year, my dad remarried that summer and my mom had gotten cancer. There was a tremendous amount of anger, pain that I didn't deal with. I have two brothers and a sister, and everyone took a different route to deal with all of these things that were happening—trauma, basically. And I was always an internal person. My sister is quicker and lets out anger, and I didn't do that as much. So I just sort of went into my own little world, I think. And then that world became totally about about eating and weight.

Kim used food to express her anger and sadness. She acted out her personal trauma through her body, and adjusted her food intake as a way to exert some power over her own small corner of the world. In a way, she appropriated our culture's mandate that women "look thin" and turned it into a drastic coping mechanism.

Unlike Kim and Linda, the majority of women in my sample used disorderly eating primarily as a way to maintain a culturally correct body image. Such eating behavior, risky though it may be, is considered "normal."

The Family and the Thinness Message: Mothers, Fathers, and Siblings

The family is a child's first interpreter of the larger world. Some families repeat the cultural values of thinness, others modify the message. Barbara's case is an extreme example of how parents can amplify the message that to be beautiful is to be loved. This message dominated her outlook on life, and contributed to the development of a full-blown eating disorder.

Barbara's Story

Barbara was about to turn 20 when I interviewed her for this book. I had known her for almost two years and I'd had an ongoing dialogue with her concerning her weight and body image, and the chronic problems she had with anorexia and bulimia. From appearances, Barbara seemed a happy, well-adjusted college co-ed. She was not overweight, but she did want to lose a few pounds and spent considerable time working out at the college gym. Her hidden history of anorexia started in seventh grade. Her bulimic symptoms began in the ninth grade and continued throughout her high school and college years.

Barbara's parents had had serious marital problems for a long time. Her father, toward whom she felt a great deal of ambivalence, had very high standards of feminine beauty. She grew up observing how difficult it was for her mother to live up to her father's expectations of the ideal woman.

> For my father, a woman has to look perfect. She has no brains. My mother has to go to my dad's functions and she has to just sit there with a smile on her face and look great at parties. My father loves it that his wife looks so much younger than everybody else . . . I don't think they were ever friends. They were just kind of physically attracted to each other. She does everything to please my father. She would go on a diet for my father. She colors her hair for my father. She got fake contacts for my father. She lies out in the sun all summer. That's all my dad wants to do, be as tan as he can, and she wants to be as tan as she can for him . . . And, oh my god, my father would get in fights, would not even talk to my mother for like a week, because her toenails weren't painted and she was wearing open-toed shoes!

Barbara did not escape her father's criticism of her own body. As a pre-adolescent she was taller than the other girls in her class and this made her feel "big." "When I was little my dad always used to make fun of me. I was never fat, just tall, but he used to pinch my stomach and say 'Barbie, you got a little rubber tire in there.' "

So Barbara stopped eating in the seventh grade. "I lost so much weight they were going to send me to a hospital, because I refused to eat. I wanted to be thin and I loved it. I ate the minimum, a little bowl of cereal for breakfast. I wouldn't eat a dessert. I remember my father forcing me to eat a bowl of ice cream. I was crying and he said, 'You're going to eat this, you know,' which was funny because he always used to call me fat. I used to lie down every night on my bed and love to see how my hip bones would stick out so much.

"When my father said, 'You're even skinnier than your sister,' I was so happy inside. It was like an accomplishment, finally for once in my life I was thinner than my sister. I remember going shopping with her to get jeans. I tried on size zero and they fell off. It was the best feeling I'd ever had in my entire life. I went back to school weighing under 90 pounds and I was about 5 feet 6 inches. I loved competitive sports, so when I couldn't play tennis anymore because I was fainting, I started eating. I started noticing that I could eat so much and get on the scale and I wouldn't even gain any weight because I was playing so much tennis. My eating was normal during that period, the eighth grade. But I wouldn't eat in front of anyone."

Then she started bingeing and vomiting. "It was awful. It was the worst feeling. You know you are about to throw up but you have to get the last bite in. I don't understand how, when you are going to throw up, you're walking right to the bathroom and you're still shoving food in. My bingeing would only happen when I was alone, before my parents came home."

Barbara's eating problems continued into her college years. She described a typical binge:

> I still binge and I always do the exact same thing. I put on my backpack and go to the local food store. I don't want to talk to anybody. I always get cookies, cake, and ice cream because it is easy to throw up. Chipwiches, brownies, sundaes. Once I stole from the cafeteria because I was too embarrassed to buy it. The minute I get back to my room, I lock the door and turn on the music. I can't throw up in my bathroom because other girls will hear so I turn on the music and throw up in my room. I'll get a garbage bag from downstairs first. It's so gross. After I throw up I feel awful—it can be so exhausting all you do is fall asleep.

For Barbara, there was no escaping the pressure to be thin and attractive "because that's what my father thinks and likes. I guess I want to live up to his standard." Yet she knew how devastating this has been for her mother and how rocky a relationship her parents had.

"I always said to my mom that I'd never want to marry someone like Dad. I don't want what happened to my mom to happen to me. He just wants my mother to look young for the rest of her life. He doesn't want her to go gray. The big joke in my house now is that my mother's going through menopause and she just cries all the time. And my dad is like, 'You're so old'. And my mother is just devastated. She looks a lot younger, but my dad always tells her she looks old. She doesn't look old. A lot of people ask if she's my sister. And I mean, she has to wear bikinis. She doesn't want to. She always has to wear full face makeup on the beach, because you can't show like any blemishes, or anything. You have to look perfect. It always drove my father crazy that my sister and I don't wear makeup.

"My father is definitely there when we go shopping. He always looks through women's fashion magazines, cuts out photos for us. 'I think you should get this outfit, this outfit.' I mean we have piles of these stupid pictures. When my mom and dad came up to visit me at college, I had to change to go out to dinner, because I wasn't wearing a skirt. I thought I looked fine, but he was embarrassed.

"I get angry, but then again, it's the way I've always been brought up. My dad would say, 'Yeah, we might be kind of crazy, but look how much we've given you.' I was always angry at my mom for never saying anything. She always knew I was right, but she would never say anything. She was always such a wimp.

"My dad has never seen divorce as an alternative. He thinks, you get married, it's for life. I know inside he loves my mother more than anybody. It's weird. He can't show it, but I know he does. When I was growing up I remember always listening to them fight. When my mother would be crying, I would say, 'Why don't you just leave him?' and my mother's reason was, 'I like my financial life style. I like going to Europe every year. I like having a summer house. I like having my summers off. If I get divorced I can't have any of that.' "

Part of Barbara's response to these pressures was bulimia. She used compulsive eating to numb her anxiety and anger, and purging to relieve her dread of being fat and unloved.

She had also begun to develop some of the psychological symptoms that are classic for women with eating problems, like maturity fears. In many ways Barbara was afraid of growing up and facing what her mother experienced as an adult married woman. She could see how women are devalued in society. As Nancy Chodorow says, "The flight from womanhood is not a flight from uncertainty about feminine identity but from knowledge about it."[13] Instead, she loved playing the kid role, making up fantasy tales and playing on the swings in the playground. As she put it, "I was always the goof ball. You know, like I never grew up. The night before I left for college my mother said

I didn't have to go. She was like, "You're my baby, I don't care if you don't go to college."

It is clear that there are a variety of pathological issues within Barbara's family,[14] but Barbara's eating problems cannot be fully understood without an awareness of their setting. The assumptions of our culture are evident in her father's demands that his daughters and wife look thin and perfect.

It is impossible to point out any one factor in the family environment that would explain how families induct their daughters into the Cult of Thinness. Some parents go out of their way to *avoid* emphasizing body image issues with their daughters. However, the college women I interviewed were quick to point out the little ways their families passed on the cultural message.

Mothers are crucial brokers of the wider cultural norms. Some research studies note how a mother's attitudes about her own body image and eating behavior influence her daughter. Having a mother who is obsessed with being thin and who diets regularly is considered one of the risk factors for the development of an eating problem in an adolescent girl.[15]

Pamela gave me an extreme example of how a mother's concerns led her daughter to take a very dim view of her own body image and self-esteem:

> My mom would always tell me that I was very chunky. She's 5' 1" and weighs about 100 pounds, so she always fits into size two and four clothes. And here I am, I can't get fit a piece of her clothing on my elbow. I'm the fat one, her fat daughter. I don't want to be near her. She always looks good, has to have the nicest clothes. I just feel very obese next to her.

Through her chronic dieting, Betty's mother conveyed an important message to her daughter when she was growing up. Later on, in college, Betty mimicked her mother's attitudes about weight and eating issues in her own way, by becoming bulimic.

> My mom considers herself overweight. She always dieted, trying all those new, different diets, and I'd go on them with her. She tried diet books and the new rotation diet. She'll fluctuate, losing some but then she'll gain it back. Right now she's at a point where she says, 'I'd like to be thinner, but your dad and I like to eat.' And they go out to eat often and it's hard to stay on a diet. So she's coming to accept it, but I think she'd still like to lose weight.

Betty described how her mother always prepared big meals. "My brothers and father liked to eat. We always had a lot of food on the table and we usually had seconds. My grandfather would reward us for cleaning our plates. Even now I feel like I have to eat everything on my plate, even if I'm not hungry."

Not surprisingly, Betty was always very conscious of her weight and body image. "I have this little chart that I try to follow. It has your height and your build, if you're medium or small boned. It has how many calories are in anything you eat. My mom put it in my stocking at Christmas."

When she was a college freshman she gained enough weight to prompt her father's remark that she was getting a "little chubby." Her bulimia started during her sophomore year, after a Christmas party.

> I made myself throw up. I had been drinking at the party and I just ate out of control. I couldn't believe that I had eaten all that stuff. It was cookies mainly. I never thought I could throw up, and then I tried and it worked. I tried so hard to keep my weight down and I was 115 and that's where I wanted to stay. And then I'd creep back up. To get all the food out of my stomach, especially that high-calorie food, I threw up.
>
> I didn't have time to join a fitness club and I tried to run, but I didn't feel like I was exercising. I could just feel myself gaining weight and that made me nervous and I felt like if I didn't have time to exercise, then throwing up was kind of the answer. I tried dieting and I wouldn't eat for awhile, and then I would eat a lot. I can't go a whole day without eating. I think that's a major reason why I throw up too. Because I like to eat."

Trapped by two conflicting messages—"Be Thin" and "Eat"—Betty used bulimia as a solution.

Mothers can also serve to modulate cultural norms of thinness and alleviate some of the pressures young girls may feel. Joanna, who is not overweight, described her mother's attitude.

> My mother, all she wants is that I'm happy. I can weigh 500 pounds as long as I'm happy. Her focus was always on my health, not so much with my appearance. So her comments were more towards that positive support. Very rarely do I remember her giving anything like negative comments about how I looked. It was encouraging. My mother would stay stuff like 'You have a beautiful face, you have beautiful hands.' She'd focus on individual qualities about me.

For the most part, the fathers of the women in my study were silent about their daughters' body image. Barbara's father (above) was unusual. So was Marsha's—her father wanted her to be thin and have perfect grades. "He always looked at us as showpieces. That's how he thought other people saw us." Her mother was very overweight when she was growing up; it was Marsha's father who told her she was fat. "I was three years old when my dad started calling me fat. He used to sing me a song, 'I don't want her, you can have her, she's too fat for me,' and I used to cry every time. It was terrible. He was repulsed by my weight, just as he was by my mother's."

Jessica's father was more typical: "I can't recall my father saying much about my appearance. Like if I put on a skirt he'd say: 'Oh, you *do* have legs' and he joked about it. Never anything negative."

Helen's father was also quiet. On special occasions he might say a word or two: "It was one of those things where if you bought a new dress for a prom, and you tried it on for him he'd say, 'Oh, that looks great' or 'very nice' or whatever. But as far as day to day, he'd never say anything. To tell you the truth I don't think he would even know what to say because that's not the type of question that he'd answer. He'd think, 'What kind of a question is that, 'How do I look?' '"

The role of siblings as mediators of cultural values has even less documentation. It is clear, however, that older siblings can be an important influence on how young girls think about their bodies.

Judith's older sister made fun of her weight. "My older sister teased me because I was larger than all my siblings. I was three or four inches taller than even my older sister. And I think I was also bigger. I started developing earlier. She used to say 'You're a fatso.' It used to get me mad, but then when I looked at myself, I would think, 'But, I'm not fat!' "

Sibling rivalry led Katy's family to label each other according to certain body features, and hers was her weight.

> One of my brothers was little, and we called him Little Bit. Jeffrey had buck teeth, Judy had freckles. My other sister had a funny nose. My older sister we called the Prima Donna because she's always putting everything on and looks so nice. Everyone had a label. My five brothers and sisters would make pig noises around me. At the time it really hurt my feelings because I'm a very sensitive person. I really was upset but my parents really didn't know I was. It bothered me that they never really said "Katy may be heavy, but she's a person, she's your sister, don't talk to her like that." They never said that and I think I resent it to this day.

As girls mature and move beyond the family, peer groups and the school environment begin to take on important mediating roles in fostering the Cult of Thinness.

The College Environment: Breeding Ground for Food and Weight Obsession

Nearly all the women I studied felt that college life had a profound impact on their body image and eating patterns. They are not alone: colleges and universities across the country are reporting dramatic increases in eating disorders among their students. Several studies on anorexia nervosa in college populations report that it affects between 6% and 25% of female students.[16] The prevalence of bulimia ranges between 1% and 19%[17] and between 23% to nearly 85% engage in binge eating.[18] Although the disorders are listed as separate syndromes, some overlap of symptoms exists. While self-starvation is specific to anorexia, obsessive concern with weight and reliance on extreme diet measures are common to both, and bulimic episodes are frequent among anorexics. A new term, "bulimarexia," has also been offered for the mix of symptoms.[19]

There are several reasons why the college environment may be a breeding ground for weight obsession and the development of eating problems.[20] First, college campuses are middle- and upper-middle-class enclaves—a group that places a high premium on thinness in women.[21] With the spread of mass media, especially television and magazine advertising, the Cult of Thinness as a prominent cultural message has reached across the class/race spectrum. As more and more women from working-class or ethnic backgrounds populate a wide range of colleges and universities, they become increasingly vulnerable to this message.

Second, college life provides a "semi-closed" environment and this tends to amplify sociocultural pressures.[22] Disorderly eating can spread through imitation, competition, or solidarity. Some researchers note that bulimia is an "acquired pattern of behavior," or suggest that there is a "social contagion" phenomenon that occurs in such an environment.[23]

Another important factor is the importance of physical appearance in dating. Students who live in close quarters, such as dorms or sororities, often feel pressured to live up to group standards of beauty, even it means engaging in eating disordered behavior.[24] It follows that schools that emphasize the dating scene may also have higher rates of bulimia.[25]

A recent study of female college students followed a group of 23 women at two universities, one predominantly white and the other historically and predominantly black. Over time, these women "shifted more and more of their interests and energy away from college work and toward the peer group, with its emphasis on romance." The authors of the study concluded that the peer-group culture was of primary importance to college-age women. This culture centered around how to meet and attract the opposite sex, what the researchers term getting a grade on the "sexual auction block." How attractive a woman could be to men depended on looking good and maintaining a thin body. *This was the primary determination of her status within a peer group.* Academic studies were not highly valued by peer group culture and did not play an important role in the day-to-day college life of these women.[26]

College Eating, the "Freshman 10," and the Vicious Cycle

The stress and strain of college, from academics to social life, magnify female students' problems with weight because women often use food as a means of calming or coping. Weight gained during college is especially detrimental in a climate already primed to value thinness. Extra pounds lead to efforts to lose, and trigger a new round of disorderly eating. This becomes a vicious cycle: stress leads to overeating, which leads to weight gain, which leads to restricted eating or overexercising, which may lead to more stress if the diet or exercise doesn't work. Even if it does work, just worrying about how to keep the weight off can trigger more eating, more weight gain, and so on.

The transition from high school to college may be a common time of weight gain for women students, as well as a period when they report that they began to diet or binge eat.[27] The "freshman 10," the ten pounds a freshman is expected to gain the first year, is a cultural prescription in league with the "pig out" and the "frat barf."[28] In my sample, students reported weight gains of 15 or more pounds. Some students reported a smaller weight gain, and only a few said they lost weight.[29]

The very fact of being away from home affects the regularity of eating patterns. Without parents to organize meals and supervise nutritional intake, students find themselves too busy to eat, or eating more than usual.

Charlotte said:

> You kind of forget about eating and then all of a sudden you gorge yourself. At home you just realize, okay, I'm gonna have breakfast, lunch and dinner. So you really don't think about food. In college I have classes on this day so I can't eat until this time. It's just different because you're used to a set pattern at home and Mom usually cooks it but here you have to get it yourself so it's very erratic. I definitely overeat.

The scheduling of cafeteria meals at college often conflicts with students' preferred eating times. Some students remarked that breakfast and dinner times were too early. Students often skipped breakfast and some had "double dinners." Julianna's remarks were typical: "At home, we didn't eat until 6 or 6:30. If you're eating dinner early and you're still up at 11, you're hungry. That's why we would order out a lot. And my roommate ate junk food. I was never a big potato chip eater or any of that kind of stuff, but she was, and having food around meant I would eat it. I gained weight quickly."

Shana told me that during her first year of college eating was *the* social thing to do. "I think actually freshman year everyone I knew gained weight. Probably at least 10 pounds. It's just like eating was the thing to do. It was a very social type thing. I can remember sending out for pizza at midnight a lot of times with a whole bunch of people and stuff; things that you wouldn't usually do."

And June said: "I was smoking all first semester of freshman year and I was gaining 25 pounds. But now I think, I won't eat that: I'll have a cigarette. And I don't binge like I used to. Every once in a while I'll binge. But the weirdest thing was when I had lost all my weight after sophomore year, and I was looking anorexic. Just to save myself, I would diet all week. Then Sunday was my day to eat everything in sight. And I'd do it to the point that I got sick. Then I wouldn't eat on Monday, and I'd diet, diet, diet until Sunday."

Melissa noted the pressures she felt from the academic environment and sometimes she used food to cope with her stress. Her bulimic symptoms were often triggered by procrastination on a paper or exam. She recounted the story of a bulimic episode that occurred one late night in her dorm room:

> I was in my room and I had to write this paper. It was due the next day. I had done my research on it. I had taken all my notes. All I had to do was to write the thing. You know, it wasn't a long paper, I knew what I was writing, but for some reason, I did not want to write this paper. I thought, maybe I just won't show up. Maybe I'll go to the infirmary in the morning. I don't know how many dumb things went through my head. I do this for almost every single test I have. I was sitting there in front of this computer and I did not finish the paper until 11:00. I just sat there, like 'I can't do this!' That night I binged on two pieces of cake, and two cookies. The minute I got up to my room, I locked the door, binged on the food and threw up. I finished the paper, and I didn't sleep at all. I went on the Stairmaster machine the next morning for 45 minutes.

Taking tests can create tension within the whole dorm. Candace told me, "There is a lot of stressful eating in college. That's the thing to do. 'Oh, study week, we are going to have to cram, oh, let's stay in and eat.' And we all did that."

Ann described the typical binge session:

> You'd be sitting around, you'd skip dinner because you were studying or it just didn't look good or whatever the reason would be. And somebody would've gone down to the store, so it would just be a matter of having to starting munching on a few potato chips. And then we'd say 'Oh, I've got a bag of something else' and pull those out. And the next thing, you'd say 'Well this is crap, why don't we just get a pizza.' But then when the pizza came you'd be on something else. There was always junk there to be eating.

The type of disordered eating these students were engaged in was a way to cope with the stress. One girl told me: "Food wasn't much of a problem in high school. Food became a psychological problem in college, and I would be bingeing and purging, because my boyfriend did something to me, because I had an exam the next day, because I felt really lonely at college, all by myself. It was psychological issue."

Usually a binge would create a fear of weight gain among the women, and be followed by group purging or individual dieting or exercise.

Sara said: "I'd eat like a slob for a week and then the following week I'd stop eating all together. A binge for me is eating some type of 3 meals a day and then when I'm home I eat everything in sight. And it's not just eating dinner. It's eating maybe two or three of everything that's there. And on weekends, I just keep eating and eating. The following week I won't eat breakfast or lunch and all I'll eat for dinner is some broccoli or maybe I'll make an ice cream sundae. I don't care too much about nutrition."

Dawn noted how even vomiting can occur in group situations: "I remember my roommates all pigged out and then they all tried to make themselves throw up. They just all went into the bathroom and threw up. And I couldn't do it. I tried but it wouldn't happen. I thought, 'This is stupid!' So I never did it again."

Long-Term Impact

What remains unexplored by researchers is the long-range impact of college life on women's body image and eating patterns. The transition from high school to college may be a particularly vulnerable time for women as they struggle with independence, growing up, and self-concept issues, and this vulnerability is related to the onset of disorderly eating. [30] It is a time when students are entering the dating/mating game more seriously, as many of them expect to meet a partner during their college years. Several female students I interviewed told painful stories about feeling unattractive and unsuccessful in finding a boyfriend; one young woman related the following story.

> I liked this guy and we were friends and then I came out one day and made the stupid mistake of telling him. I said 'You must know how I feel about you,' and then he said 'You'll never be Cybill Sheppard.' At the time I was horrified. Now I can laugh, because he's certainly no Bruce Willis. But I couldn't say this because I was so hurt at the time and so his comment really did a thing on my psyche. And in a matter of 18 weeks I lost 35 pounds.

I conducted several follow-up studies of women from their senior years in high school to their senior years in college. They reported downward shifts in their self-esteem in terms of physical attractiveness, social self-confidence, assertiveness, and popularity, especially with the opposite sex. I also found a connection between this decline and the development of more disturbed eating patterns. [31] In fact, those women whose eating patterns remained problematic throughout the four years of college showed a pattern of diminishing self-concept over time compared to those whose eating patterns were normal or whose eating issues improved.

For example, Miriam said: "My self-esteem just went down. I went from a public school to this private college, where it seemed to me like everybody was really rich, and everybody had been to private schools. I'd been so protected before, when I surrounded myself with people like myself. I just went off the deep end."

Irene told me: "I had a horrible experience that first year of college because of where I lived, who I lived with and the alienation I felt. I gained weight, of course. I was living with three girls who had their own food issues, as everyone I have lived with seems to have had. On top of that, you have homework that probably wasn't being completed because you're thinking about your weight. Eating became, you know, how to get through the day."

There is no simple cause-and-effect relationship here, and it is difficult to generalize the findings from one study. More research could explore the reasons for these self-esteem declines during college and how this is specifically related to eating issues. One picture that begins to emerge from such research so far is that college transitions serve to reproduce traditional gender stereotypes of women, creating an opportunity for the Cult of Thinness to flourish. Some researchers have speculated that the co-educational college environment may act as a "null environment"[32] for its female students—one where women are neither encouraged nor discouraged in the pursuit of any but the most traditional of roles. In light of their previous socialization experiences, the lack of encouragement alone may have some devastating consequences for the developing self. As one researcher writes, "Even though men and women are presumably exposed to a common liberal arts curriculum and other educational programs during the undergraduate years, it would seem that these programs serve more to preserve, rather than to reduce, stereotypic differences between men and women in behavior, personality, aspirations, and achievement."[33]

Understanding the Cultural Context of Eating Disorders

In summary, disorderly eating is not a sign of psychopathology, but a strategy that is a "normal" part of the female existence. While there may be evidence of psychological trauma, or even biochemical deficits associated with some of the eating problems occurring in the young college women I have described, these psychological and biological theories fail to address the more general issues. The link between the cultural norms of thinness and the individual is mediated by the family, school, and peer group. They translate and embellish society's messages.

Yet it would not be enough, for example, to get Barbara's father to be more aware of the impact of his "stay young and beautiful" message on his wife and daughters, or to counsel the family to deal with their emotional issues. That would not resolve Barbara's problem with her body image and eating disorders. Her problem is wider than that, reinforced at every turn by the world she lives in.

8 New Recruits for the Cult of Thinness: Young Girls, Men, and Ethnic Women

"When I see these twigs of people in the magazines and on TV, I say 'I'm going to go on a diet.' I think I won't look good in those clothes because I'm not that thin. You almost want to get thin just so you can wear the right clothes. I watch my junior high friends—they look like something out of a magazine."—Darcey, age twelve

"There is an increase in men having cosmetic surgery . . . There is nothing more symbolic of age and infirmity and the loss of masculine vigor that for a 50- or 60-year-old male to develop fatty deposits in his breasts, so that he can't go the local country club pool and bare his chest. Those men are in here having their breasts or abdomens suctioned."—Boston plastic surgeon

"All the models are perfect all-American with blond hair and blue eyes, and they're all tall and skinny."—Michele, age fifteen

Marketing the Cult of Thinness to Pre-Teen Girls

The average fashion model is white, 5 feet 10 inches tall, and weighs 110 pounds. She is approximately 30 pounds less than the average American teenage girl, and almost six inches taller.[1] Her looks are relatively rare, yet her image is so pervasive that it is difficult for girls to see themselves as anything but "wrong" in comparison.

Young girls face an endless barrage of messages from beauty magazines and television, and from classmates and parents and doctors, that thinness is valued and obesity is a liability. Many of them, by virtue of being female, white, and middle class, are already primed to join the Cult of Thinness. There is growing evidence that increasing numbers of new recruits, at ever-younger ages, are coming from this population. A recent research study suggests that even pre-adolescents are joining the diet craze and some are stunting their growth as a result.[2] Why are some of them so fearful of fat?

As we have seen earlier, there are profits to be made from convincing certain vulnerable groups, like young people, that they need to purchase goods and services to feel good about their bodies. More and more, advertising campaigns for fashion and beauty products are targeting children.[3] These industries are well aware of the increased purchasing power of pre-teens. One researcher explains: "Today's parents spend more money on their children than any prior generation; children and adolescents have unprecedented amounts of money at their disposal, which they spend on fashion, beauty, and entertainment or leisure products; and children have gained increasing power in a wide range of purchasing decisions made by their parents."[4]

Many of these products are directly aimed at promoting body insecurity. (Am I fat? Does my hair lack body? Do I have blemishes?)

At a developmental point in their lives where they are sensitive to peer group pressure and media messages, kids seek out "how to" messages. As "guidelines about how to behave, young adolescents may be particularly susceptible to popular media stereotypes, especially those values and ideas presented by entertainment and fashion industries as vital elements of 'youth culture.' "[5]

Teen magazines provide a powerful voice for defining what it is young girls are supposed to be doing with their lives; what is recognized as important; what is valued. It is here especially that they are socialized to take up the obsession with their bodies, even during their time in school. The overall message is that if you want to be beautiful and happy and to get a boyfriend, then you need to look like the models.

One young woman I interviewed told me: "Magazines were the big thing, especially in the teen years of your life. There were always articles talking about how to become thinner and sexier and how to attract the opposite sex. Everything seemed to focus on that in the media."

In a 1993 back-to-school issue of *Teen* magazine, a ten-page advertisement on "Boy Appeal," carries an obvious message. The first page begins, "School's back in session and these girls are ready for the books—and the boys—this semester! Follow these friends from their morning makeup to classes and dates and find out what they do to look great while they're on the go!"

The ad depicts a cluster of girls who are sitting on the school steps talking to a group of boys. In the background are their "Caboodles"™ (plastic carrying cases) filled with a variety of products: makeup, contact lenses, hair spray, and so on. Throughout the spread, the copy and photos describe how young girls can work in their beauty routines before, during, and after school.

For example:

Kristi mists on Tribe just as she leaves for school in the morning and keeps it with her in her locker during the day for extra spritzes— especially when she bumps into that really cute new guy in school! Tribe's blend of roses, lilies, jasmine and orange blossoms makes Kristi hard to resist!

The last page, showing Mary and Holli coming home from school, reads:

Classes are out (finally!), so Mary and Holli are on their way to meet the boys for an after-school date. They carry their Caboodles everywhere—Mary's Signature travel case gets straight A's in style and design, slim enough to carry with plenty of room to hold all of her favorite makeup. Holli chooses the Caboodles Signature with two swing trays and mirror. It gives her the flexibility to carry everything to take her from morning to night—she puts her life into it!

Getting straight A's in school may not compare with getting a Caboodle travel case that gets straight A's for style and design. In fact, in one ad the message is clear that girls should be careful that hitting the books doesn't distract them from boys:

Jamison has been eyeing Marisa in class for days, but her eyes are on her books. When he finally works up the nerve to ask her for a date, Marisa removes her glasses before she says "Yes!" Marisa goes from daytime to nighttime "frame free" with Bausch and Lomb Medalist™ contact lenses for the prettiest eyes of the evening!

The ad even gives Marisa tips for convincing her parents that she needs to buy contact lenses.

"Thin promises" are also a big part of the teen magazine world. Their messages link thinness with love and happiness, often solely in terms of having the right body to attract the opposite sex. Diet and weight loss products fill the mail order section of these magazines, like this ad for New Shape™.

There's a girl in every school who is so beautiful—so foxy—that all the boys just can't keep their eyes off her. A girl who is surrounded by cute guys wherever she goes. A girl that can have any guy she wants. Very soon that girl could be you! Now you too, can have the kind of body that boys like and girls envy: thin, shapely legs and thighs, firm behind, narrow waist, and foxy, exciting curves in all the right places. New Shape™ does more than just help you get rid of excess fat. It helps you add shape and firmness too. It's a total shape-up system that you can customize to your own special needs. . . . You can change your flabby embarrassing features—but keep your good parts.

Not only are media images in teen magazines depicting the slender ideal, but educational materials are also mirroring the trend. A 1992 study of third grade textbooks illustrations showed that in each decade since 1900, girls' body image in these textbooks became thinner. The study focused on those illustrations depicting the entire body of the child, where the gender could be clearly identified. There was no significant trend for the boys' body image.[6]

Fear of Fat in Pre-Teens

The medical establishment may have helped trigger an excessive fear of childhood fat by casting doubt on the old image of a chubby, healthy baby. A late 1960s study on feeding patterns in rats found that the number of fat cells the body produced was influenced by overfeeding early in life.[7] Doctors and nutritionists applied these findings to humans and told parents that over-feeding their infants and toddlers could place them at risk for long term weight problems.[8] Historian Roberta Seid says:

> These findings seemed to explain why obese children so often grew into obese adults and why so many Americans couldn't control their appetites. Their excess fat cells wouldn't let them. This heartening theory implied that there was a long-term cure for America's leading health problem. Fat adults still might be difficult to treat but the new generation could be spared the agonies of its overweight parents."[9]

In due course, researchers found that it was not valid to extrapolate the findings from rats to humans, but not before the professional community of pediatricians[10] and the popular press spread the fear of fat to parents with young children.[11] As Seid notes, this fear created a new market, spawning a range of weight-loss books for younger children as well as the development of kids' weight-loss camps.[12]

The "fat camp" for pre-teens and teens is a phenomenon akin to adult weight loss programs. The promotional literature from these camps shows how well they understand that, for kids, the social consequence of being fat is *the* primary issue. "Erica [15 years old], 215 pounds at 5 foot 4, has checked into Camp Camelot because she hated being fat. She hated being taunted, being called "fatso" and "lardo." She hated looking in the mirror, hated what she saw there so much that she would punch her pillowy face in a search for cheekbones and scream at herself inside her head: 'I HATE YOU! I HATE YOU! YOU ARE SO FAT!' "

After losing 25 pounds at camp, "Erica has lost more weight . . . Her classmates 'really freaked out' when they saw her at school because 'I am a totally new person.' She got a new haircut and new makeup and is hoping for a boyfriend 'who will be with me forever.' "

The reward of thinness is social acceptance, "a new image—a new you," and better feelings about one's "slimmer, trimmer self." But by binding self-esteem so closely to weight and physical appearance, this attitude may also set the stage for a psychologically damaging cycle of weight gain and self-hatred. The youngster who loses weight at summer camp and "improves her self-esteem" is in danger of feeling worthless if she regains some weight when she returns home.

Studies show how early the cultural mirror begins to distort girls' perceptions of their body size and weight—by six to nine years of age children have a definite preference for particular body shapes and sizes.[13] One study asked children to ascribe attributes to three different body silhouettes—thin, normal, and fat. The results noted that "children of both sexes and all body builds express strong preferences for athletic or lean body builds and dislike of chubby or heavy builds." In addition, these body build preferences were associated with "peer nominations, interpersonal relationships, and prestige among children and adolescents."[14] Research also showed an exaggerated concern with obesity among young children and adolescents. One study conducted in 1986 on boys and girls 9 to 18 years old, from middle-income families, revealed a startling finding: over 50% of the young girls characterized themselves as fat, while only 15% would be considered overweight according to national height and weight data.[15] In this same study, 31% of the nine-year old girls thought they were too heavy now or feared they would become fat later, 81% of ten year-old girls reported that they were on a diet.[16] Among other age groups, the number of dieters ranged from 46% to 89%.[17] Other research shows that girls as young as five or six years old may fear gaining weight.[18] Lauren, in one of my college interviews, recalled:

> I was big for my age and they called me Baby Huey, you know, after the fat cartoon duck. That was always my impression of myself—a real clod. At six years old I can remember feeling big. It was horrible, because in ballet class I wanted to be like the other girls, petite and pretty. I look at those pictures now, and I looked fine. I was a beautiful little girl, but at the time I didn't feel that way.

An article in a 1988 issue of *Pediatrics* reinforces Lauren's observation:

> . . . adolescent girls and women generally have a conception of attractiveness that does not conform to biologic reality. In other words, the usual standards for healthy bodies are based on actual measurements of thousands of individuals. Ranges of statistically and clinically defined normality are developed. What many adolescent girls are telling us by their behavior is that they accept neither the biological reality nor the medical concept of appropriate weight.[19]

While there have been some reported cases of pre-teen anorexia nervosa, eating disorders are not widespread in this younger population. However, severe dieting practices are quite common among pre-teen girls.[20] There have been numerous research studies[21] documenting the "fear of fat" among young girls who are of normal size. This fear exists despite their knowledge of nutrition and their own "normal" body weight. Over 50% of the underweight adolescents in one study described themselves as extremely fearful of being fat and said that they did not apply their basic nutritional awareness to their eating habits.[22] There is also some evidence that young girls who practice extreme dieting risk "nutritional dwarfing," where poor eating causes short stature and delayed puberty.[23] *Newsweek* reported that 13 year olds were following such regimens as the "Jell-O Diet," and the "Popcorn Diet."[24] In fact, self-imposed malnutrition has become a health concern in clinical pediatric practice.[25]

While the mass media provides young children with images of the culturally desirable body, the role of family members and peers is also very important in how young girls acquire a positive or negative self-image (as we saw in "Becoming a Certain Body"). My interviews with young women included many references to the criticism or negative comments of siblings, friends, and parents (especially fathers). In some cases it had a lasting effect on their self-esteem.

Donna told me, "My dad used to call me fat. I wouldn't have noticed on my own. It was hard for me, growing up, being a fat kid. Dad hated it. He wanted me to be thin."

Young boys, Donna's peers, also sent her a message about her body. She lived through a special trauma in seventh grade.

> I was going out with a guy who was very cute and I was feeling like the happiest person in the world. His friends started giving him a hard time because I was fat and also the smartest girl in the school and you don't go out with the fattest and smartest girl in the school. So he broke up with me—even though we had confessed love to each other. To this day, he was the only guy I ever felt that way for and he dumped me, totally unexpectedly. It was real hard because not only did he dump me but he convinced all his friends to give me a hard time too, I guess so it didn't look like it was just him. So again, I was totally ostracized, this time for being smart and not just for being fat."

"Why My Body Was Doing That": The Onset of Puberty

Some researchers suggest that young girls' problems with weight, body image and eating are linked to the onset of puberty, which brings a 20 to 30% increase in body fat.[26] Though it is critical to maturity and reproduction,[27] many young teenagers regard this normal increase with horror. Many of my college interview subjects recalled this time in their lives with pain or embarrassment.

> I got my period in the 10th grade. I thought something was wrong with me. I kind of knew what it was, but no one really sat down and explained to me why my body was doing that. I felt ashamed at first.

I was twelve when I started getting a chest. I hated it. My sisters weren't developing yet—even my older sister. I guess at first, I thought it was fat. I never wanted it.

Peer-group comparisons become especially important now, as bodies begin to change and develop. Many women I interviewed remarked that while they did not feel like they were "one of the boys" during this time, they certainly did not feel like one of the girls. Some went out of their way to tell me they were simply girls who wanted to do boy things.

Mary remembers:

When I was a kid I used to love hanging out with guys, much more than with the girls. You know, if they wanted to look at dolls or whatever, I would say, 'I want to go have some fun.' So I'd go out and I'd be on the swings. I remember one day my mother came to school. I guess she got a call from a teacher. She came and yanked me off the slide, because I was playing with the guys. She told me never to play with them again. I asked her why, and she said, "Just because."

She used to put me in dresses for parties, especially when it was my own party. I'd kick and scream, but she'd put them on me. I 'd go back into my room, change into my favorite pants and by that time I knew that she couldn't do anything about it, because if she did it would make a scene and she didn't like that, especially at my party.

Being a tomboy or denying a feminine identity for a while protects some young girls from the pressures of the female role. It relieves them from being attentive to fashion, body image and getting caught up in "boy appeal." For most of them it is a phase; only a short delay before they succumb to what's involved in being female in this society.

Men and Their Bodies

Men, especially younger men, appear unbothered by media images of what their bodies should be. In higher education and in the workplace, as well as in personal relationships. Jim and Ken are two college-age men I interviewed. Here's their response to the question, "Do you think women are more weight-conscious than men?"

JIM: "Women are more weight-conscious. Men have never had to worry—the only time I think men get really weight-conscious is when they're playing sports. When I swim, I have to be conscious of it, but otherwise I never think of my weight. When I'm not in swim season, I'll just grab a candy bar and not think about it. I don't expect to marry because some woman thinks I look good. I am not going to try and look a certain way because I'm going to get married. Hell no."

KEN: "Traditionally men are more active in athletics, so weight watching is not a problem. The media and models have placed a lot of pressure on women to be painfully thin as opposed to the strong and muscular man."

The stakes for being physically attractive are lower for men than for women. Sociologist Marcia Millman, who interviewed many overweight women and men, put it this way:

> In our culture, being fat more deeply affects a woman's self-image, her social identity, and her treatment by others . . . Even men who are 200 pounds over-weight by conventional standards told me that being fat wasn't very important in their lives; they seemed not to think about it very much and claimed it didn't cause them suffering in work or in their personal relationships. [28]

In this study of obese men and women, Millman noted that men did not "psychologize" about how they looked and did not see their weight as a personal or emotional problem. What was important to the men she inter-viewed were the health consequences of getting fat. [29] However, complex social forces are changing the picture for men.

The Rise of Male Consumers

As all Americans are becoming more consumer-oriented rather than produc-tion-oriented, "men as well as women will be evaluated increasingly in terms of how they measure up to media images of attractiveness rather than their achievement in work." [30] A recent study of articles in men's magazines over a twelve-year period notes "a statistical trend for an increase in weight-loss focus" and suggests that men are becoming more appearance conscious. [31]

There is a huge financial potential in promoting body obsession and anxiety in men, and it is no wonder that within recent years the market for men's body products has grown dramatically. The diet and cosmetics indus-tries have developed marketing strategies that prey on men's weight and appearance insecurities. Certain diet soft drinks and weight loss products are targeting the male market. [32] Estee Lauder, Inc., and Elizabeth Arden, compa-nies widely known for their women's cosmetics, have been offering a wide range of men's skin care items for years.

Says one observer,

> . . . over the last few years there has been an increase in the number of men used in advertising and an increase in magazines designed specifically for the male reader. Stereotypically good-looking men now advertise milk, cars, food and a multitude of consumables. Men's magazines are full of tall, dark, handsome men modeling clothes and even the male perfume market has taken off. [33]

Body Dissatisfaction Among Males

The market emphasis may be working, because there is growing evidence that men are becoming increasingly dissatisfied with their weight and body image. Data on this comes from two studies conducted in 1972 and 1987: "13% of [both] men and women were dissatisfied with their height in 1972, but this rose to 20% and 17% respectively in 1987. About 10% were dissatisfied with their face, but this doubled to 20% in 1987. Weight was even more problematic: 35 and 48% of men and women were unhappy about it in 1972, but this increased to 41 and 55% in 1987."[34] What these studies reveal is that there are high levels of body dissatisfaction among *both* men and women. Even though our culture still exhorts a large percentage of women to look thin and attractive, men are also targeted.[35]

This growing emphasis on attractiveness in both men and women is amplified by the medical establishment, which links a trim body with health:

> Western society currently places an unprecedented emphasis on life-style change and self-management as the major health-promoting activities. The burden of illness has shifted from infectious diseases to cardiovascular disorders, automobile accidents, and cancers, many of which are considered preventable though behavior change. Looking healthy is the external manifestation of the desired healthy state, so the body symbolizes the extent of one's self-corrective behavior.[36]

Another important reason for men's increasing attention to appearance is the changing nature of gender roles. Traditionally, "The woman is supposed to be attracted to the man for his social achievements (wealth and power) and simply because he is a man, not because of any special effort on his part to make himself attractive to her."[37] Today, as women are gaining some economic resources and positions of authority, they are starting to shift the balance of power within society. Noted psychologist and eating disorder specialist Judith Rodin writes:

> Men's appearance concerns also seem affected by shifting gender roles and expectations. Once a man could be assured of his masculinity by virtue of his occupation, his interests, or certain personality characteristic. According to historian Mark Gerzon in his book, *A Choice of Heroes: The Changing Faces of American Manhood* there have been five traditional archetypes of masculinity throughout history: soldier, frontiersman, expert, breadwinner, and lord. Frontiersman and lord are no longer available roles for anyone, and expert and breadwinner are no longer exclusively male. Men may be grasping for the soldier archetype—the strong, muscle-armored body—in an exaggerated, unconscious attempt to incorporate what possible options remain of the male images they have held since youth.[38]

Under the Knife

Body dissatisfaction has lead some men to take drastic measures. While women still outnumber men seven to one in terms of undergoing some form of cosmetic surgery, men are beginning to opt for this route to good looks. The *Utne Reader* reports that "It is estimated that the number of male procedures has increased by 50 percent over the past four years. Some prominent plastic surgery centers report that men represent as much as 30 percent of their business."[39] As the general population ages, men who were once young and secure about their body images may be subject to fears of devaluation not unlike those of women.

I asked a successful plastic surgeon for his views on why women continue to account for the majority of those seeking cosmetic surgery, and the reason for the increase in the statistics on these procedures for men. He told me:

> We still don't view men and women as the same, and until our aesthetic demands on men equal the ones that continue to be placed on women, we will still see more women responding to those pressures through cosmetic surgery. However, I do see an increase in men having cosmetic surgery, and the most common reason that I'm encountering is economic. It's the male who is rising up the economic ladder and usually achieves his greatest economic success in his fifties and sixties. Now he needs a physical appearance that is consistent with his power and his place in society. He must stay physically trim, must not have loose skin, must not have anything that suggests infirmity. His looks maintain his power because we're all being judged by the vigor of our appearance. This is more true for men, because our society continues to economically reward them over women. Even though a 50 year-old woman entering my office might give the same reasons as a man— she wants to maintain a high position in her career and feels that without a youthful appearance, she would lose out—I think for most women it's an issue of self-esteem. With aging, they have diminished self-esteem and they want to regain something that they feel they've lost."

He went on to say: "Men and women choose different cosmetic procedures. Women primarily go from the waist down; they go for hips, thighs and buttocks—even knees. I never met a man who asked to have his thighs suctioned. Men are obsessed with their paunches and the fatty deposits on their chests."

Dieting and Eating Disorders in Men

Roberta Seid notes that, historically, "Men could not easily be sucked in to dieting because of the persistent belief that a big, strong body was masculine and sexy. Even if it wasn't too strong, a big body gave the illusion of power and sexual vigor."[40] While most men who are dissatisfied with their weight deal with it through exercise,[41] there is evidence that others are taking to dieting in increasing numbers. "Already diet soft drinks, light beers, and other diet products are being marketed by male movie stars and athletes. Men are finally getting hooked into feeling immoral if they eat the wrong foods." Researchers suggest that if this trend continues, "the next ten years will see an explosion of weight problems in males."[42]

Given these trends, we can imagine the Cult of Thinness spreading to the male population. Most research data shows that eating disorders in men are still relatively rare compared with women, with the ratio of female to males approximately 9 to 1.[43] However, "experts now believe that in the 1990s many more boys and men are suffering from these problems than doctors or the public realize."[44] Judith Rodin notes how men try to suppress their concern with physical appearance. "Perhaps they try to keep their body image concerns a secret. It's less socially acceptable for men to think and worry about their appearance than it is for women to do so. Men experience their body concerns as unmasculine, and therefore embarrassing and shameful."[45]

According to one psychologist, "Bulimia may be more prevalent among males than we thought since it's an easy disease to hide and men are reluctant to come in for treatment."[46]

Neglect and misdiagnosis may plague many cases of male eating disorders, as "Patrick's" story demonstrates, reported in a 1993 issue of The Wall Street Journal.[47] Patrick spent nine lonely years "alternating between starving, gorging himself and exercising three to four hours a day. It took nearly that long for his family to talk about it at home, much less to acknowledge to others that he suffered from anorexia and bulimia. His coworkers at the . . . bank where he worked also never caught on, even though he would lock himself in the office bathroom and do calisthenics to work off lunch. Sometimes he would run up and down 12 flights of stairs in his business suit. No one at the bank said anything to him about how he looked and acted . . . Some of his overweight colleagues, unaware of his private agony would say, 'I'd give anything to have your body.' "[48] Even when Patrick entered group therapy where he was the lone male, "the women either refused to recognize his problem or berated him for intruding."[49] Patrick eventually quit his banking job to concentrate on recovering from his eating disorder. He looks back on pictures of himself when he was very thin and now sees "a very frail, old-looking person."[50]

It is clear that males are increasingly being drawn into the pursuit of attractiveness. But this does not signify any wholesale conscription into the Cult of Thinness. A general urge toward extremism in this issue may not exist in men. For example, evidence shows that men are far less concerned with dieting than women are. In one study of college age men and women "67% of the college women surveyed were dieting at least some of the time whereas only 25% of the men were dieting."[51]

Men tend to equate the term "heavy" with muscularity: "Men care a great deal about their body build and they aspire to a widely held ideal of physical attractiveness, the muscular mesomorph."[52] Part of the male/female, mind/ body dichotomy relates to men's needs to be powerful and dominant. Some researchers note that "We believe that the muscular mesomorph is the ideal because it is intimately tied to cultural views of masculinity and the male sex role, which prescribes that men can be powerful, strong, efficacious—even domineering and destructive."[53]

Male Subgroups

The greatest number of converts to the pursuit of muscular thinness occur in particular subgroups. Men who are heavily involved in those athletic pursuits requiring weight norms (e.g., wrestling, horse racing) may be more at risk for eating problems.[54] Some studies hint that gay males may be at even greater risk because of the importance of appearance (from body build to clothing) in their culture.[55] David Crawford in his book, *Easing the Ache: Gay Men Recovering from Compulsive Disorders* highlights the importance of physical attractiveness within gay society: "To some degree, we can identify with the image women have had thrust on them. Seeing ourselves in abject terms of physical attractiveness, we—like many women—are extraordinarily self-conscious about our looks."[56] Some research also suggests there is greater body dissatisfaction among gay males. A research study comparing heterosexual and homosexual college men notes: "gay men expressed greater dissatisfaction with body build, waist, biceps, arms, and stomach than did heterosexual men. Homosexual men also indicated a greater discrepancy between their actual and ideal body shapes than did heterosexual men, and showed higher scores on measures of eating regulation, and food and weight preoccupation."[57]

The emphasis on thin appearance (ectomorphic) within gay society may be changing with the increasing incidence of AIDS in the gay community and its devastation on the body. Instead, some researchers suggest that a more muscularly powerful (mesomorphic) body image may be emerging to "avoid the appearance of illness with AIDS. The illness has been described in slang usage as 'slims' in some countries."[58]

The Spread of the Cult of Thinness to Other Social Classes, Races, and Cultures

The Cult of Thinness primarily occurs in wealthy Western societies among white, upper-middle-class, educated females.[59] It is their stories that have comprised this book. The excessive pursuit of thinness has been rare among people of color in the United States (e.g., Blacks and Latinos)[60] and in non-Western developing societies such as Asia, Africa, and South America.[61] In fact, these societies view obesity quite positively. As psychologist Esther Rothblum writes, "In developing countries, the major causes of death are malnutrition and infectious disease, and thinness is unlikely to be viewed with envy, rather, increased body weight is associated with health and wealth."[62] Furnham and Baguma also note: "to a person living in a poor country an obese body may be considered a healthy body for two reasons: first, fat deposits laid down mean that people may survive 'lean' periods more effectively; second, because one has to be fairly wealthy to afford food and could equally use this wealth to acquire medical treatment."[63]

I found an in-depth look at the role of obesity in the day-to-day lives of ethnic women in a 1989 study entitled *Que Gordita!* conducted by Emily Bradley Massara.[64] She wanted to know the cultural factors that contributed to the weight gain among a small sample of Puerto Ricans living in one Philadelphia neighborhood. Massara investigated the life histories of several first-generation women who "have a distinct sense of social identity as Puerto Ricans"[65] and who were defined as "medically obese." All were married and had children. In her research she found that particular cultural definitions of overweight were important factors in the high degree of medical obesity she found among her sample. Obesity within this community was not considered a sign of illness. Instead, heaviness indicated "tranquillity, good appetite and health."[66] Thinness was feared. To be thin meant to be malnourished and diseased. In fact, when given a series of photos of body types (thin to obese) to rank order in terms of attractiveness, the women in her sample gave a narrower range of acceptable thin weights, but a wider range of acceptable heavier weights.

Food plays an integral role in Puerto Rican life. As Massara notes, "one of the ways in which the 'good wife' and mother expresses her love for her husband and children is by presenting family members with large helpings of food and manifesting concern over amounts of food eaten."[67]

Weight gain was an expected part of a woman's social life, especially upon marriage. It was a "sign, particularly to her family, that she was adequately provided for."[68] Women who lost weight got negative reactions. One woman whose weight declined from 170 to 140 pounds when she divorced her husband reported: "When I lost weight, people said: 'You're so skinny! What happened to you?' So many people told me fat didn't look bad and I look better that way because I had a shape."[69]

Massara observes that

Linguistic terms, such as "pretty little plump one" (*gordita buena*), reinforce the notion that a certain degree of heaviness in women is positively valued. The plump (*gordita*) woman may also be referred to as a "total woman" (*mujer entera*) because she is considered to have a "beautiful body [shape]" and good health. . . . "How plump!" (*!Que gordita!*) is one expression which suggests shapeliness and health and is used in a highly complimentary manner.[70]

In Massara's sample, the men appeared more sensitive to their weight than the women. She notes that this discrepancy stems in part from the expectations associated with the provider role: "Both men and women expressed a belief that men should, as one informant explained, "be in shape for the work they do," or, "they shouldn't let themselves go.""[71]

Massara found that the men were more likely to diet and were more concerned about their appearance. One man in her sample lamented, "Already I'm 'over the hill'! The girls like someone who is nice and slim.""[72]

Massara also observed that American values of female thinness were beginning to appear within this group of first generation women, as the process of acculturation spread to their children, the second generation.[73] "For instance, many mothers showed an awareness of Western medical concepts about the dangers of heaviness in children. With regard to adult weight, some children encourage their mothers to reduce so that they will look more attractive. By the same token, some of the more acculturated women actively practiced eating restraint.""[74]

What is happening in one Puerto Rican neighborhood is a microcosm of what is happening throughout non-affluent classes and racial groups within American society, as well as within non-Western societies as a whole.[75]

Thinness and Black Culture

Research studies conducted in the late 1980s and 1990s confirm that the Cult of Thinness is spreading beyond the white middle class. Eating disorders have reportedly increased among the American black population. One researcher speculated that "increasing affluence among some blacks, and thus their access to traditional white middle class values, and the homogenization of life style and priorities, perhaps as a result of the increasing influence of the media, have finally penetrated the black culture: the young black female (and perhaps the male) is getting fatter and is becoming more concerned about her fatness."[76] The problem appears particularly acute among persons of color who are upwardly mobile. A case study of the development of anorexia nervosa in seven middle class black and Hispanic adolescent women marked the following factors:

They encountered early our society's conviction that thinness and trimness are the essential ingredients that lead to success. Thus, these girls, who were already feeling different and suffering form a low self-esteem and a powerful need to be accepted, sought integration with society through rigid dieting and an extremist adoption of the current societal standard of slimness.[77]

Statistics reveal that, in general, the rate of obesity among black women is greater than among white women. Research indicates that black women are less concerned with being thin than white women[78] and eating disorders among black women are rarely reported in the clinical literature.[79]

Gladys Jennings, Associate Professor of Food Science and Human Nutrition at Washington State University, says, "there's a cultural standard from our African heritage that allows for more voluptuousness and padding on black women."[80]

Traditionally, the African American community's ideal has been more realistic and sensual: "Women's bodies were substantial. They had breasts and hips and curves and softness."[81] Black psychologist Marva Styles notes that black women have maintained their cultural roots and strong bonds through their relationship with soul food: "The essence of black culture has been handed down through oral history, generation after generation in the African tradition, through the selection and preparation of soul food. The determination to hold on to native foods by bringing seeds into this country may be symbolic of the ever-present determination to preserve the African culture through food."[82]

That preparation became a primary definer of a black woman's sense of herself: "the Black woman gains a sense of pride as she watches her extended family—her man, her children, and maybe her grandparents, sisters, nieces and friends—enjoy the soulful tastes and textures prepared by her skillful hands."[83]

As in the Puerto Rican community, plumpness is a sign of health and prosperity And it is a signal to the black woman that she is doing a good job. Styles remarks on the generational changes that have taken place in her attitude toward food compared with her mother's. "Slimness, however, is not valued by middle aged and older black women. My mom worries about my slimness because at 5' 4" , I barely weigh 120, and I am a middle-aged woman. She asks me often, 'Are you eating properly these days?' Maintaining my weight at 120 pounds is hard for me, because I was taught to enjoy eating and preparing food. If I ate the kind of food my mom prepares consistently, I would probably weigh 150 by now. . . . Staying slim is difficult in a culture that values cooking and eating."[84]

But it is not just the maintenance of food customs that explains the high obesity rates among both black and Puerto Rican communities. Women of color face double jeopardy in that they are subject to racial as well as sexual discrimination. If we include discrimination based on social class, they can be said to suffer from "triple jeopardy."[85] In the interviews I conducted with women of color, I found that they used food as a nurturing mechanism to cope with oppressive social and economic conditions. Bingeing on large quantities of food is a "cheap" way to find temporary relief from conditions such as sexual abuse, poverty, racism, and sexism. The intake of large quantities of food in a short time period can serve to numb, soothe, and literally "shield" (with fat) some women from physical and emotional trauma. This type of bingeing or compulsive eating, which leads to obesity, is an eating disorder. We noted eating as a coping mechanism for those white women in my sample who had experienced sexual abuse, where the drive for thinness was secondary to the use of food to deal with their trauma. Preference for this form of bingeing is a perfectly rational means of dealing with the pain that blacks and others experience living in our society. For these reasons, the Cult of Thinness will most likely spread to the black population only as black women begin to gain status in terms of upward mobility. A recent study[86] reports that black college-age women have "less drive to achieve thinness" and their weight loss strategies are "realistic and less extreme." However, those black students who "endorse attitudes reflecting rejection of their black identity and idealization of white identity" were found to be more at risk for the development of eating problems and disorders.[87]

It is important to point out that the Cult of Thinness has a different context for a woman of color, who does not have the standard Anglo-Saxon features in terms of hair texture, skin color, and overall build. One black sociologist writes: "white feminists who write about body image, such as Naomi Wolf, often fail to acknowledge the particular concerns that black women face because of the combination of racism and the beauty myth."[88]

She lists the important appearance issues that came up in discussions with other black women:

> African women are subject to the same pressures to attain an ideal of beauty as are white women in North American society, but efforts to approach the blonde, thin, young ideal are made at an even greater cost for black women. Weight preoccupation is not a central concern for many black women,[89] but weight is one among many factors that preclude black women from attaining 'beauty' according to the cultural archetype. Three issues came up again and again when I talked with other black women: skin colour, hair texture, and body size.[90]

As one black researcher says, the definition of white beauty is dependent on the denigration of what is not white: "Blue-eyed, blond, thin white women could not be considered beautiful without the Other—black women with classical African features of dark skin, broad noses, full lips and kinky hair."[91]

Capitalism is helping to spread (white) Western values across racial, class, and ethnic lines. These values are being transmitted via satellite to television sets across all cultures, regardless of race, class, and level of industrial advancement. Developing societies import Western norms of beauty through the purchase and consumption of Western media, clothing styles, and beauty products. Increasingly, non-Western societies will be presented with an "ideal" of feminine beauty that takes on Anglo-Saxon traits. As non-Western women attempt to meet the ideal, they may deny the very features that give them their racial and ethnic identities—and their unique beauty.

9 Breaking Free from the Cult of Thinness

"I think it's a state of the heart. I don't think there's anything in particular that, for example, the fitness industry or the diet industry can do because that's going from the outside in. It has to be a question of where we've come to inside."—Fitness trainer, Boston area fitness club

"I don't think social change happens from the inside out. I don't think people have inner children somewhere inside waiting to be nurtured, re-parented, and their natural goodness released into the world . . . our inner selves are constructed by the social and political contexts in which we live, and if we want to alter people's behavior it is far more effective to change the environment than to psychologise individuals."—Celia Kitzinger[1]

"The Personal is Political."[2]—Carol Hanisch

Does change come from "inside out" or do we have to change society to change individuals? Can one escape the Cult of Thinness by means of self-help books and programs, and various therapies? Or is what has happened to women's bodies a political issue, inseparable from our personal concern?

From the Inside Out: Self Help Books and Self-Help Treatments

Nancy, a college sophomore, commented: "I think weight is important, but I think a person's correct weight really depends on the person herself. I think it's wrong to say 'I've got to look this certain weight' for anyone else but yourself. If you can feel comfortable the way you are then it will make you a more positive person I think—make you a better person."

Nancy's advice about "feeling comfortable with the person you are" echoes the philosophy of the self-help literature. Self-help books of the 1980s and 1990s have replaced the role of consciousness-raising groups of the feminist movement in the 1970s, except that now all the talk is about personal issues with little analysis of the "personal as political."[3]

The prevailing message in this literature is that you can change your state of mind by adopting the "right" attitude. One very popular book entitled *Making Peace with Food*[4] is intended as a resource for those women who are weight preoccupied, are dissatisfied with their body image, or have an eating disorder. Its central notion is that each individual needs to be more accepting of her body image and more at peace with food. The author, Susan Kano, points out the risks of dieting and the tendency of the body to want to reach its "natural" weight. She offers a series of exercises for finding a more positive body image and a better sense of self, as well as for dealing with stress and anxiety. *Bodylove*,[5] another book in this genre, emphasizes how to develop body acceptance while improving one's image through a range of beauty techniques, such as the artful use of cosmetics. *Transforming Body Image*[6] relies on exercises that draw on the imagination to help women love the body they have.

In preaching self-acceptance and presenting ways of reclaiming one's self-esteem, these books can assist individual women in overcoming body insecurity and hatred. They also help women to "tap into a community of sorts; they 'feel less alone' when they read."[7] As one analyst of self-help therapy notes: "They do some of what the consciousness-raising movement did twenty years ago: They let us share our deepest, perhaps most shameful pain with people in the same boat, and they provide examples of how others have extricated themselves from similar situations."[8]

Gloria Steinem's *Revolution from Within* was written, she says, because after a "previous dozen years working on external barriers to women's equality, [I] had to admit there were internal ones too."[9] Her book suggests that many women's self-esteem has been damaged by abuse or some type of deprivation. In this view, rejecting the Cult of Thinness begins with recovery and reclaiming your "Inner Child," the authentic self. The research of Carol Gilligan and others shows that around the age of 11 to 14 girls begin to lose their authenticity, the childhood selves that are confident and self-assured. They "go underground," begin sentences with "I don't know, but. . . ," become more attuned to what others think of them, and lose faith in their own opinions. Their self-esteem diminishes, and depression among this age group increases. They start to worry about things such as weight and how they look to others, especially boys. They become concerned about competitiveness within their peer group, and their academic performance, especially in math and science, may begin to decline.[10] The "revolution within" refers to the process of regaining what women lost in their pre-teen years. As one therapist working with eating-disordered women says, "young teenagers encounter enormous conflicts when joining the cultural definitions of adult womanhood, in which they are forced to risk their authentic self in order to fit into androcentric norms, values and images. For many girls, going underground begins with their bodies as they struggle to fit into culturally defined molds."[11]

To get a sense of what life is like for pre-teens on the brink of losing their "authentic" selves, I interviewed three nine year-old girls: Beth, Monica, and Willa. Beth and Monica have sisters who are five years older. Willa is the oldest in her family. I asked them to tell me what it is like to be nine years old, and what they did with each other in their free time. They remarked, "just playing and having a lot of fun hanging around." When I asked them to elaborate they chorused, "We play with Barbie dolls." Beth has nearly two dozen Barbie dolls in her collection. (First she told me she had only 12, but her older sister, who overheard this part of our conversation, reported that in fact she had 23.) I asked the girls what type of games they played with the Barbies. They replied in unison: "We play the Mean Sisters."

The Mean Sisters game is a reenactment of the type of identity conflicts pre-pubescent girls go through. Beth put it this way when she talked about her older sister.

> When you get older you start liking boys, wanting to be skinny and you want to get dates and stuff like that. When my sister Amelia was young she just got dressed, washed up, combed her hair, ate her breakfast and left for school. Now it is different. She gets out of bed, she washes up, then she starts fiddling with her hair and making it into ponytails and seeing what she likes, then she dresses up for a really long time 'cause she wants to make sure she looks really, really good, then finally she goes down and eats breakfast. Probably it will happen to me. It's natural. Because as you get older you like boys."

As Beth and Monica stand on the brink of adolescence, they feel that what has happened to their older sisters is "naturally" going to happen to them, and they are worried. Beth said: "I'm afraid that if I'm very beautiful and I wore high heels and stuff then I couldn't go out and play sports anymore. I'd be putting on make-up and fixing my hair, if I end up like my sister."

They want to hang onto the selves they have, but also realize that as they get older they will need to confront what their older sisters have become. In their play, they act out their ambivalence.

> We split the Barbies into groups. (We have Ken dolls also.) First there is the mother. We usually don't have a father. We pretend he is away on vacation or got divorced or died. We split the mother's children into little kids and big sisters. The Skipper [Barbie's younger sister] dolls are the little kids. They don't have big boobs. Then we divide the Barbies into mean sisters and nice sisters. The nice sisters treat the little kids nicely and the mean sisters usually try to kill the nice sisters because they want to have all the power and the nice sisters prevent that. The mean sisters don't want the Skipper kids nagging them and stuff.

I asked these pre-teens what the nice girls usually talk about. They replied, "They talk about opera and dancing. The mean girls talk about boys, plans to kill the little sisters and they talk about being beautiful. They talk about big breasts, their beautiful eyes and hair."

"What made the sisters mean?" I asked. They said, "They only care about how beautiful they are, they don't care about anybody else, they are snob sisters, they like to get all the power. The nice sisters don't care about how they look. They just want the boys to like them for their selves and personality."

At the height of the game the bad girls kill off all the nice girls and the little kids. (Apparently the mother has gone out when this takes place.) They sometimes kill them by stabbing or drowning the nice sisters and kids. When the good girls are dead, the bad girls have a celebration. "Like a party, where they jump up and down, and make a lot of noise, and usually get sent to their rooms by their mother. Once all the good girls are dead the bad girls use their power on boys. The boys like the mean sisters because they are beautiful and they are rich. They spend their money on good food. They get their money from the nice sisters' allowances, which they are saving up for a college education. The bad girls sneak out in the middle of the night and walk out with boys."

There is revenge, however. The nice sisters and little kids come back either from the grave or from the hospital. The mother then abandons the bad sisters (sends them to their rooms). And the game ends.

For now, Willa, Beth, and Monica are letting their dolls bear the burden of society's expectations.

Strengthening Self and Spirituality

Feminist writers offer other avenues to strengthen women's selfhood, from forging stronger mother-daughter bonds to developing one's spiritual life. In the hopeful vision of *The Mother-Daughter Revolution: From Betrayal to Power*, mothers and their daughters form a new kind of alliance. Its authors believe that mothers' influence is of primary importance in shaping their daughters' future lives. But they need to recover their own lost selves—the ones that went underground at adolescence in the face of patriarchal culture. By finding and reintegrating their truthful, assertive childhood voices, mothers can help prevent the same thing from happening to their vulnerable daughters. They can initiate a process of validation for young girls. The authors claim: "Not only does an authorizing mother validate her daughter's reality, but she adds her authority as a mother, as a woman who has experience in this culture, to amplify and harmonize with her daughter."[12] The ultimate goal is to have a community alliance of mothers and daughters and discerning males who, together, would resist the devaluation of women's selfhood.

bell hooks, a noted black feminist, stresses the importance of spirituality in "becoming your own person." She feels that it is imperative to move the individual from being an "object" to being a "subject":

> There is such perfect union between the spiritual quest for awareness, enlighten-
> ment, self-realization, and the struggle of oppressed people, colonized people to
> change our circumstance, to resist—to move from object to subject; much of
> what has to be restored in us before we can make meaningful organized protest is
> an integrity of being. In a society such as ours it is in spiritual experience that one
> finds a ready place to establish such integrity."[13]

For bell hooks, recovery of the self through spirituality[14] is a central requirement for social change.

The Role of Therapy

In Chapter 1 we mentioned depression, anxiety, and dysfunctional families[15] as oft-cited factors in eating disorders. A number of "disease" models have been used to explain why women join the Cult of Thinness. One is the addiction model, which labels women's eating and body image problems as illnesses. Like the views expressed in the self-help literature, this model of behavior places responsibility for the cure of the disease with the individual, assisted by medical treatment, psychotherapy, or self-help programs. Overeaters Anonymous, one of the largest self-help groups for women with weight and eating problems, is based on the same principles as the original quasi-religious twelve-step program, Alcoholics Anonymous. The core assumption is personal accountability. Recovery comes with accepting one's lack of power over eating issues and embracing a belief in a "higher power." Twelve-step organizations take pride in their apolitical character; in fact they go out of their way to avoid identification with any political group. Instead of looking at women's empowerment as a solution to the problem, they suggest that women admit "powerlessness" over their "disease." The emphasis on the individual totally ignores the social, cultural, political, and economic context of women's problems with weight obsession.

Although they are free and, therefore, outside the market economy, twelve-step programs still fuel the medicalization of women's obsession with weight. They replace a focus on women's oppression and exploitation with the apolitical perspective of addiction, suggesting that therapy, rather than political action, will provide the "cure." Women in these programs are locked into depending on others, who repeat such messages of powerlessness as:

- You have no power over what has happened to you and you are to admit that you are powerless.

- You are to trust a higher power to help you overcome your addiction.

- You alone are responsible for curing yourself. Self-discipline, determination, and trust in a higher power are your tools for recovery.

- We are not interested in understanding the economic, political, and sociocultural context that has led you to your addiction, or what wider social economic and political forces might need to be changed.

Some therapy programs *can* help women recover from an eating disorder. They alleviate stress and can help individual women cope with their body and food issues. Treatment may range from a hospital stay, to drugs, to individual therapy, group therapy, family therapy, or hypnotherapy. But the focus is primarily on the individual or family unit, with little emphasis on wider cultural factors.

There are also a number of feminist therapies available, which begin to make the links between womens' eating issues and cultural oppression.[16] Some feminist therapists who support a broader social perspective are conducting self-help groups that offer some alternatives to the continued obsession with patriarchal standards of beauty. Such groups encourage women to give up perpetual diets and examine the meaning of food for them. Other therapists have applied a philosophy known as "fat liberation." Their goal is to change the individual's attitude by helping her feel good about her present body weight.[17]

Perhaps the self-help market was created in part by feminism's inability to provide solutions for attaining "the longing for personal transformation that its successes have awakened in women."[18] For example, in the late 1960s and 1970s, feminist health activists promoted the idea of self-help. In this context it meant that women learned and shared information about a range of issues, from body image to women's social and economic conditions. Self-recovery inspired women to question their own conditions and to take action in dealing with them.

Today, therapists and self-recovery authors do not attempt to change the "outside," but resort to individualistic "inside" treatments. Since the purpose of self-help books and therapeutic intervention is to treat the eating disorder or the body image disturbance, they rarely look at the causes connecting the individual with society. These treatments, in effect, do not challenge existing structures of domination and power in our culture. Instead, they tacitly accept these oppressions, including the pressures on women to look attractive, and even address ways in which women can strive to attain a more culturally accepted body image. There is also a prevailing attitude that women who put a great deal of their energy into their own growth and development are "hurtful and destructive to others." As psychologist Harriet G. Lerner notes,

Women tend to feel so guilty and anxious about any joyful assertion of self in the face of patriarchal injunctions that each small move out from under is invariably accompanied by some unconscious act of apology and penance. . . . Recovery then, to my mind, is a sort of compromise solution. It teaches women to move in the direction of "more self" while it sanitizes and makes change safe, because the dominant group culture (never fond of "those angry women") is not threatened by sick women meeting together to get well.[19]

Individual solutions may be the only recourse for those who, because of their age, gender, race, or class, are not in a position to effect societal change. In fact, for some, giving up the pursuit of thinness may be unwise. As we have seen, women are rewarded or punished for having the culturally correct or incorrect body image.

Social Change from the Outside In

If we recognize the social, political, and economic forces that help to support and sustain the Cult of Thinness, then it is clear that simply helping individual women deal with their weight and body image will not resolve these issues on a broad scale. Every day, what we eat, what we wear, and how we view our bodies are very much shaped by the outside forces of our broader cultural setting. These forces are not only associated with institutions like government or the education system, but are central components in how we perceive our personal lives.[20] A different way to escape the Cult of Thinness focuses on the structural features of society rather than on the individual. These solutions aim to change the climate in which the Cult of Thinness flourishes, and chop at its very roots.

Addressing the Political Economy of the Cult of Thinness Through Social Activism

In this book we have examined the sociocultural and political-economic framework that supports the Cult of Thinness. We observed how capitalistic interests, such as the diet, cosmetic, beauty, and health industries, and the mass media benefit from promoting women's body insecurity. An ultra-slender body ideal also helps controlling patriarchal interests, since it requires women to divert money, time, and energy away from more empowering activities.

To address these issues demands a critical examination of the current structural features of capitalism and patriarchy—even a boycott of those industries involved. One activist strategy involves organizing women at the grass-roots level to target and boycott consumer goods whose advertising is offensive to women's body image. Boston-area women who are against the use of overly thin models in advertising formed a group known as Boycott Anorexic Marketing (BAM). Their purpose is to "curtail the practice of featuring waif-like, wafer-thin models in ads for a variety of products by identifying companies considered to be culprits and asking consumers not to buy their wares." As the founder of the group explained: "So many women in this group felt powerless at the way our culture applauds anorexia and we thought of this boycott as a way to talk back."[21] They successfully targeted a Diet Sprite advertisement that depicted a very thin model who, the copy told us, as a teenager was nicknamed "Skeleton." Their effort prompted the sponsor to retract the commercial. Making women aware that they have the power to change societal attitudes through their purchasing decisions could diminish the Cult of Thinness.

The tobacco industry is another important target for boycotts. Over the last 25 years, the tobacco industry has created a female market by feminizing certain brands of cigarettes such as Virginia Slims and Capri. Their ads strongly suggest smoking as a way to help women lose weight and imply that women who quit cigarettes will gain weight. Even though cigarette smoking accounts for about one-quarter of the cancer deaths in women each year, a growing number of teenage girls are taking up smoking.[22]

We must also look into the controlling role of the medical industry. Currently, anorexia and bulimia are classified as illnesses, requiring medical or psychotherapeutic interventions. Eating disorder clinics promise to be big money makers. One might question whether labeling an eating disorder as primarily a medical problem serves the economic interests of the drug companies, who have developed an antidepressant drug (fluoxetine) for bulimic women. Their theory is that a chemical imbalance in the brain is the root cause of bulimia and other eating disorders in women.[23] Yet a current review of the link between depression and bulimia finds "inadequate support" for this relationship.[24] Drug companies and medical researchers who are eager to explain women's problems with food as depression, and to market a profitable "cure for eating problems," should have their motives examined.

Social activism within our own communities is another route to social change. There are many opportunities to become involved in challenging the Cult of Thinness. I asked the nine year-old Barbie owners about what could be done to change older girls' concerns about how they look. They immediately made the connection between their toys, their body image fears, and the greater population of children. They wanted to take action. Beth said, "I would like to change the Barbie doll's shape. I'd like to make her normal, not big boobed, just like a normal person." Monica decided to write a letter to the Mattell toy company.

Figure 11 Capri Super Slims Cigarettes, December, 1994

Dear Mattell:

Make your Barbie dolls less adorable or stop selling them because little kids and older kids and even younger sisters and brothers play with Barbie dolls and in one second they will want to look as skinny as them. So either stop selling them or make them more normal, a little chubbier, not as big boobed and not standing on their tippy toes. Make more pants, and not only shirts with puffy shoulders which come down to their belly button. Make full bathing suits not only bikini bathing suits.

Some women advocate combining our enthusiasm for getting our bodies in shape with social activism. Body work for women tends to be an isolating activity. As one commentator says,

. . . we work out alone and devote attention strictly to ourselves. Concerted group action seems foreign to exercise. While camaraderie may spur one on to more repetitions or laps, rarely does it promote friendships or genuine teamwork. This solitude and self-inspection tunnels the vision and leads one away from any investigation of the social sources of flab, stress and heartache.[25]

Women I talked with suggested organizing a 10K walk/run in which women of all sizes and shapes were encouraged to participate, to challenge cultural attitudes regarding ultra-thinness. Others mentioned a one-day bike ride to promote awareness of eating disorders and to advocate insurance coverage for the treatment of these problems.

Social Activism Through Education

Public education is also important. One educational program called EDAW, Inc. (Eating Disorders Awareness Week) is sponsored by the medical community (physicians, dietitians, mental health professionals, etc.) as well as by educators, coaches, and athletes. The purpose of this week is to raise public awareness regarding the prevention and treatment of eating disorders through promoting a healthy attitude toward body image. The organization has challenged cultural pressures toward thinness (especially those pressures from the multibillion-dollar diet and fashion industries) and links these pressures to the outbreak of eating problems. It challenges cultural standards of beauty with the idea that "it's what's inside that counts."

Educational programs on eating disorders are beginning to show up in school curricula. In one private school in Boston, all eighth- and ninth-grade girls are required to take a ten-week course on anorexia, bulimia, and compulsive overeating.[26] At Harvard University, the hot line known as Echo (Eating Concerns Hotline and Outreach) provides a forum for men and women who have eating issues or are concerned about a friend who has an eating problem. They also sponsor speakers and films to educate the wider community. The Center for the Study of Anorexia and Bulimia has put together curriculum materials for grades seven through ten dealing with such topics as emotions and eating, body image, and the cultural pressures on women.[27]

Several national associations are also educating the population at large. The National Association to Aid Fat Americans was founded in 1969. It provides its members with "tools to be happy with yourself as you are, to fight discrimination and stereotyping and to understand the effects that dieting can have on the body and mind." The association has gone out of its way to fight fat discrimination, from launching a write-in campaign against greeting cards offensive to the obese to educating physicians about the problems of undertaking surgery to cure obesity.[28] The National Association of Anorexia Nervosa and Associated Disorders is another self-help and educational organization that aims to promote healthy body image and eating attitudes.

A New Vision of Femininity: Breaking Down the Mind/body Dichotomy

Western society's prescription of what it means to be feminine needs a rewrite. Current definitions of femininity are dictated by a social system that gains control over women by defining them primarily in terms of their bodies. The split between mind and body is central to our idea of what it means to be male and female—and in our culture, mind is valued over body. But dichotomous thinking is a powerful mechanism of social control and oppression. It separates groups into "we" and "they," instead of allowing a diversity of images to flourish.

In order to break down the mind/body dichotomy, our institutions must change. They must acknowledge and reward a broader view of "the feminine" that includes aspects of both mind and body. Truly eradicating the Cult of Thinness will ultimately require women to become politically active in changing basic institutions—educational, economic, family, legal, political and religious. Women must challenge the industries that feed on body insecurity. They need to change the messages girls and women absorb from families, schools, and jobs—all places where women are rewarded or punished daily for being in the "right" or "wrong" body.

To effect change in the economic arena for example, women must continue to demand equal opportunity in the workplace. Sexual harassment, ageism, and weightism must be included with issues of salary equity and job promotion. Our legal institutions must address gender discrimination as well as the growing violence against women that continues to undermine feminine power and authority. In education, it is important to close the gender gap in mathematics and science, and to build awareness of gender bias within and outside the classroom. This will require teachers and counselors to change male-based teaching environments and provide positive encouragement to female students to overcome years of anxiety about taking advanced math and science courses.

In the sense that the body is a cultural construction, perhaps we can begin to build another culture—one that offers alternatives to Madison Avenue messages. In the social arena, we need a re-envisioning of femininity, where women look hard at current cultural norms and free themselves from these definitions. Groups of women are beginning to do this.

In *Female Sexualization: A Collective Work of Memory*, Frigga Haug describes a group of German feminists who have undertaken an exciting new project using their own bodies as objects of study.[29] The goal is to "unravel" how gender socialization has created and molded bodies over time. In this process, each woman chooses a body part—hair, legs, and so on—and then asks the others to recollect events in their lives that focused on this area. They circulate their written memories among the group, where body socialization stories are "discussed, reassessed and rewritten."[30]

Photography has been an important influence in the control and definition of feminine appearance, from advertising images to the family photo album. Another new way to re-vision femininity is photo therapy, through a technique known as "reframing." The idea behind photo therapy is that "each of us has sets of personalized archetypal images 'in memory,' images which have been produced through various photo practices—the school photo . . . is one example."[31] These photos are often "surrounded by vast chains of connotations and buried memories."[32] The reframing process[33] involves a challenge or description of "visual discourse" by parody and reversal. Through playing with photo images of themselves, women 'retake' old images and thereby gain some control over how they define their physical appearance and sense of self.

Still other strategies of re-envisioning femininity require women to become politically active, even on the smallest personal level. I interviewed a group of women ranging from their thirties to sixties and asked them what advice or solutions they might offer to the next generation. What would help younger women to promote social change around women's problems with body image? They agreed that changing society would most likely happen in small increments. It was not going to happen quickly or on a massive scale, but on a small group and peer group level. In the words of one woman in her forties, a school administrator, social change starts with our "significant others." She commented:

> A person's life extends out, like an embrace. I know I can work on myself, but I also know the impact that I have with my stepchildren and my siblings and friends. And they have an influence on me. I think we all have that sphere of influence to work on. It's all of our sisters and daughters that we need to make aware of the Cult of Thinness. I think demonstrating some of my changed ways of being and my changed attitudes toward my body have made a difference to my stepdaughter, my younger sisters, my best friend. They've watched me gain weight and not freak out about it too much. I mean, not get crazed as I would have in my 20s, where two pounds was cause for terrible alarm and self-abuse. I think by just living the way I'm living and calling their attention to it enough, it's made an impression on them.

Ellen, a writer, noted how she has tried to influence her extended family by asking them to not pay so much attention to her eight year-old daughter's looks.

My relatives can never greet one another without saying, "Hi, you lost weight! Don't you look good!" That's hello, the first thing. One time my aunt Mary came to see us and she said "Ellen! . . ." then she stopped and didn't know what to say, because I'd gained a little weight. I looked at her and felt sorry for her, she was so over-dyed, over-made-up, over-dressed. Because she's so afraid of who she is under it. I'm not terribly assertive with my family, but they know they can't pull that around my daughter. They can't talk about her figure, or what she's eating. I won't let them comment on her appearance. They can't do that, because to me, it takes her away from her childhood.

In their own ways, older women can reach out to the next generation to change the culture within which young women grow up. Maybe it happens when a teacher pays attention to how she talks about weight and body image in the classroom. Perhaps it's when a mother stops dieting to demonstrate to her daughters that she is breaking free of the cultural demands she has had to endure in "becoming a certain body." Maybe it's in a family's attitudes at the dinner table. As Miranda, one the students I interviewed, told me: "My parents never used food as a temptation or a weapon, like "you won't get dessert if you do something," or "you are being sent to bed without dinner." Food was never a reward and it was never a punishment."

These personal gestures are important examples of how social change can start within our own close circle of friends, relations, coworkers, or students.

Making a Life: Becoming Your Own Person

I set up a false dichotomy at the beginning of this chapter by suggesting social change as an either/or phenomenon—"outside in" or "inside out." Understanding the role that *both* types of change offer is fundamental to extricating young women from the Cult of Thinness and enabling them to "become their own person."

Finding the space to develop the self appears critical to the healthy maturity of young women. In fact, the process of self-making is crucial to moving beyond the Cult of Thinness. I asked Anna, the religious cult member we met earlier, what enabled her to leave the cult after twelve years. She had entered in her early twenties, gone through an arranged marriage with a fellow member, and had begun raising a child within the cult community. She told me she had begun to question her guru's rigid rules, and to reject his controlled version of reality, especially when he sent her young son away to school in India. "I reached a level of strength where I knew that it was time to go on and have a life of my own. I finally felt like I could figure out on my own who I was and what I wanted to do and that I didn't need to live inside that prison."

A heightened sense of self-esteem characterized those college-age women

I interviewed who felt they had put the Cult into perspective. They had begun to feel good about who they were as individuals. To some extent they had reclaimed the "self" of early childhood. I sensed that these young women were beginning to make a life for themselves that did not involve concentrating solely on their bodies, but encompassed both the mind and the body.

Evelyn described her ability to resist the Cult of Thinness by learning more about her feelings and focusing not on "What am I going to eat today, but what am I going to do today." She said:

> The more confidence I had in myself, the more self-esteem I got. I would say it comes with experience. Probably learning about life, knowing how I feel about certain things and really feeling sure about how I felt. The more I was sure of my feelings, the more I didn't care what anybody else thought. So then I had self-esteem because I felt good about who I was and where I was coming from. If I were 10 or 20 pounds overweight now I feel that it wouldn't bother me as much because I'm happy with the person that I am. I have a younger cousin who is very overweight and a lot of the problem with the girls that I know is that they don't like themselves. And eating is a diversion from thinking about what's going on in their lives or what they should be doing. Whereas I'm directing attention to doing things that I want to accomplish.

Jennifer is also making a life that does not buy into the mind/body dualism. She is the youngest of five siblings from a white, middle class, intact family. Jennifer's mother was in her early forties when Jennifer was born, and stayed home when she was growing up. As Jennifer puts it, "My mother worked until she was six or seven months pregnant with me. Then she quit until I was about in the sixth grade. And then my father stayed home and she went back to work. They both retired when I was in high school." Jennifer is a highly independent and self-confident young woman. She recently graduated from college, is looking for her own apartment, and is about to embark on her career. I asked her to define her "sense of self." "Everything I do, I do for me, not to please other people. If I decide to exercise I do it not because I want to lose weight. And if it were to lose weight it would not be because my boyfriend would look differently towards me. I exercise because it feels good, it releases tension."

I asked Jennifer what quality most characterized her. She replied:

> Probably that I'm my own person. I won't do something because that's what the crowd is doing. Recently I broke up with my boyfriend. We'd been going together for a while. I met him last summer, we both had a cottage in the same place. We became friends first, and then it grew into a relationship and we had a lot of good times together. And I guess, although it was never really mentioned, marriage was sort of kicked around. It was sort of like "Well, what do you want in the marriage?" and stuff like that. I discovered that he wanted a wife like what my

mother was and what his mother was. I don't want that, I want to work. A career is very important to me. He wanted a homemaker.

My ideal package would be to have a career, meet a guy and date him a year, be engaged for about a year, and then get married. All this still having my career, being married for several years so that we could be together and get to know each other and then start a family. And then I'd probably take time off, about a year or two and then go back to work. I know I would go crazy staying home.

I asked Jennifer if she sometimes felt it would be difficult to attain her "ideal package."

I can't say it hasn't crossed my mind because it has. But then again when I look and see my friends getting married and what they are getting into and I can see that they are not going to last. I have plenty of time. I don't want to be content with just being married. It's really funny, because I think I have this fear of marriage. My mother was saying, when you get married it's not the end of the line. Your life does go on. I really feel like I have to accomplish so much more. I want to do more—travel, kind of test my limits, even though I'm scared to do it because I'm scared of failure. But I'm not going to settle."

For women to "make a life" there must be space for them to forge an identity that is not bound by traditional definitions of what it means to be female. Working for a new femininity, based on integrating the mind *and* body, and creating a society that values women—that is the best antidote to the Cult of Thinness.

Notes

Introduction

1. The reported ratio of female to male anorexics and bulimics is 9 to 1. See: Richard A. Gordon, *Anorexia and Bulimia: Anatomy of a Social Epidemic* (Cambridge, MA: Basil Blackwell, 1990), p. 32.

2. A range of research studies in the United States and elsewhere (primarily in affluent Western cultures) document the increase in eating disorders from the early 1960s to the 1990s. One researcher claims that the rates of eating disorders in the general population "increased by a factor of at least two" (p. 152). While some have speculated that the rise in eating disorders may be a result of greater awareness and detection, the vast numbers of clinical research data "make such an interpretation unlikely" (p. 153). See: Richard A. Gordon, "A Sociocultural Interpretation of the Current Epidemic of Eating Disorders" in *The Eating Disorders*, eds. B.J. Blinder, B.F. Chaiting, and R. Goldstein (New York: PMA Publishing Corp., 1988), pp. 151–163.

The rates of anorexia nervosa and bulimia, while increasing, are still not high in the general population as a whole. However, these disorders are especially high in the female adolescent student population, which is thought to be at greatest risk. Some researchers estimate that in the United States one in every 200–250 women between 13 and 22 suffers from anorexia nervosa, and that between 20% to 33% of college women control their weight through vomiting, diuretics and laxatives. See: Steven Levenkron, *Treating and Overcoming Anorexia Nervosa* (New York: Warner Books, 1983), p. 1; Susan Squire, *The Slender Balance: Causes and Cures for Bulimia, Anorexia, and the Weight-Loss/Weight-Gain Seesaw* (New York: G. P. Putnam's Sons, 1983).

Research suggests that the prevalence of eating disorders continues to increase in frequency. See: Katherine A. Halmi, "Anorexia Nervosa: Demographic and Clinical Features in 94 Cases," *Psychosomatic Medicine* 36 (1974):18–25; Dolores J. Jones, Mary M. Fox, Haroutum M. Babigan, and Heidi E. Hutton, "Epidemiology of Anorexia Nervosa in Monroe County, New York: 1960–1976"; *Psychosomatic Medicine* 42 (1980):55.–558; R. E. Kendall, D. J. Hall, Anthea Hailey, and H. M. Babigan, "The Epidemiology of Anorexia Nervosa," *Psychological Medicine* 3 (1973):200–203; J. A. Sours, "Anorexia Nervosa: Nosology Diagnosis, Developmental Patterns, and Power-Control Dynamics," in *Adolescence: Psychological Perspectives*, eds. Gerald Caplan and Serge Lebovici (New York:Basic Books, 1969), pp. 185–212.

In fact, the rate of women's problems with food may be *underestimated*. My research on college-educated women reveals that eating disorders exist along a continuum from very severe cases to more mild, sub-clinical cases (while women in this category do not fulfill all the diagnostic criteria for bulimia and anorexia, they are still obsessed with weight issues, such as binge eating, fasting, and extreme dieting, etc.). These students' eating issues would not be diagnosed as abnormal. See: Sharlene Hesse-Biber, "Eating Patterns and Disorders in a College Population: Are College Women's Eating Problems a New Phenomenon?" *Sex Roles* 20 (1989): 71–89 ; and Sharlene Hesse-Biber, "Report on a Panel Longitudinal Study of College Women's Eating Patterns and Eating Disorders: Noncontinum versus Continuum Measures," *Health Care for Women International* 13 (1992): 375–391.

3. Hillel Schwartz, *Never Satisfied: A Cultural History of Diets, Fantasies and Fat* (New York: Free Press, 1986), p. 240.

4. D. M. Schwartz, M. G. Thompson, and C. L. Johnson," Anorexia Nervosa and Bulimia: The Socio-cultural Context," *International Journal of Eating Disorders* 1 (1982): 20–36. See also: A. Morris, T. Cooper, and P. J. Cooper, "The Changing Shape of Female Fashion Models," *International Journal of Eating Disorders* 8 (1989): 593–596.

5. D. M. Garner, P. E. Garfinkel, D. Schwartz, and M. Thompson, "Cultural Expectations of Thinness in Women," *Psychological Reports* 47 (1980):483–491; Claire V. Wiseman, James J. Gray, James E. Mosimann, and Anthony H. Ahrens, "Cultural Expectations of Thinness in Women: An Update," *International Journal of Eating Disorders* 11(1992):85–89.

6. Having internalized this message, modern women exert on themselves a subjection of the body equal to the institution oppression described by Michel Foucault. See M. Foucault, *Discipline and Punish: The Birth of the Prison*, translated by Alan Sheridan (New York: Pantheon Books, 1977). See also: S. Bartky, "Foucault, Femininity and the Modernization of Patriarchal Power," in *Feminism and Foucault*, I. Diamond and L. Quinby, eds. (Boston: Northeastern University Press, 1988), pp. 61–88.

7. For a description of cult behavior see: Max Weber, *The Sociology of Religion* (Boston: Beacon Press, 1963). See also: R. Stark and W. S. Bainbridge, *The Future of Religion: Secularization, Revival and Cult Formation* (Berkeley: University of California Press, 1986).

8. Sylvia Walby in her book, *Theorizing Patriarchy*, notes that the term has been used to refer to societies in which men ruled because of their position as head of the household. See: Max Weber, *The Theory of Social and Economic Organization* (New York: Free Press, 1947); C. Pateman, *The Sexual Contract* (Cambridge: Polity Press, 1988). See: S. Walby, *Theorizing Patriarchy* (Cambridge, MA: Basil Blackwell, Ltd., 1990).

9. This definition is taken from the work of Sylvia Walby. She presents one of the most comprehensive theoretical discussions of the term. See: S. Walby, *Theorizing Patriarchy* (Cambridge, MA: Basil Blackwell, Ltd., 1990), p. 20.

Chapter 1

1. Georgia Dullea, "Big Diet Doctor Is Watching You Reaching for That Nice Gooey Cake," *New York Times* (December 1, 1991): 65.

2. Natalie Allon. "Fat Is a Dirty Word: Fat as a Sociological and Social Problem" in *Recent Advances in Obesity Research*:1, ed. A. N. Howard (London: Newman Publishing, 1975), pp. 244–247. Allon 's observations on one dieting organization spans four years.

3. For a discussion of the "sacred" and "profane" see: Emile Durkheim, *The Elementary Forms of the Religious Life*, translated by Joseph Ward Swain (New York: Collier Books, 1961), pp. 52, 53.

4. Garfinkel's (1981) research supports this viewpoint. He notes that women's competitive striving for success can lead to the development of eating disorders in young women. See: P. E. Garfinkel, "Some Recent Observations on the Pathogenesis of Anorexia Nervosa," *Canadian Journal of Psychiatry* 26 (1981): 218–223. See also: D. M. Garner and P. E. Garfinkel, "The Eating Attitudes Test: An Index of the Symptoms of Anorexia Nervosa," *Psychological Medicine* 9 (1979): 273–279.

5. Pauline B. Bart, "Emotional and Social Status of the Older Woman," in *No Longer Young: The Older Woman in America. Proceedings of the 26th Annual Conference on Aging*, ed. Pauline Bart et al. (Ann Arbor: University of Michigan Institute of Gerontology, 1975), pp. 3–21; Daniel Bar-Tal and Leonard Saxe, "Physical Attractiveness and its Relationship to Sex-role Stereotyping," *Sex Roles* 2 (1976): 123–133; Peter Blumstein and Pepper W. Schwartz, *American Couples: Money, Work and Sex* (New York: William Morrow, 1983); Glen H. Elder, "Appearance and Education in Marriage Mobility," *American Sociological Review* 34 (1969):519–533; Susan Sontag, "The Double Standard of Aging," *Saturday Review* (September, 1972), pp. 29–38.

6. Alessandra Stanley, "A Softer Image for Hillary Clinton," *New York Times* (July 13, 1992): B1, B4.

7. See: T. Horvath, "Correlates of Physical Beauty in Men and Women," *Social Behavior and Personality* 7 (1979): 145–151 and T. Horvath, "Physical Attractiveness: The Influence of Selected Torso Parameters," *Archives of Sexual Behavior* 10 (1981): 21–24. See also: Sharlene Hesse-Biber, J. Downey, and A. Clayton-Matthews, "The Differential Importance of Weight Among College Men and Women," *Genetic, Social and General Psychology Monographs* 113(1987):511–528.

8. E. D. Rothblum, "The Stigma of Women's Weight: Social and Economic Realities," *Feminism and Psychology* 2 (1992):61–73.

9. J. Polivy and C. P. Herman, "Dieting and Binging: A Causal Analysis," *American Psychologist* 40 (1985):193–201.

10. Hilde Bruch, *Eating Disorders: Obesity, Anorexia and the Person Within* (New York: Basic Books, 1973).

11. Harrison G. Pope and James I. Hudson, *New Hope for Binge Eaters: Advances in the Understanding and Treatment of Bulimia* (New York: Harper & Row, 1984).

12. Marlene Boskind-Lodahl and William C. White, Jr., "The Definition and Treatment of Bulimiarexia in College Women," *Journal of the American College Health Association* 27 (1978): 84–86, 97.

13. William Ryan, *Blaming the Victim* (New York: Pantheon, 1971).

Chapter 2

1. See: Michelle Zimbalist Rosaldo, "Woman, Culture and Society: A Theoretical Overview," in *Women, Culture and Society*, eds. Michelle Rosaldo and Louise Lamphere (Palo Alto, CA: Stanford University Press, 1974), pp. 67–87. See also: Shirley B. Ortner, "Is Female to Male as Nature Is to Culture?," in *Women, Culture and Society*, eds. Michelle Zimbalist Rosaldo and Louise Lamphere (Palo Alto, CA: Stanford University Press, 1974). Both authors note that women are symbolized as closer to nature and men are more closely identified with culture. Male activities are given preference over female activities.

2. Paul Rosenkrantz, Susan Vogel, Helen Bee, and Donald Broverman, "Sex-Role Stereotypes and Self-Concepts in College Students," *Journal of Consulting and Clinical Psychology* 32 (l968): 287–291. See also: P. A. Smith and E. Midlarksy, "Empirically Derived Conceptions of Femaleness and Maleness: A Current View," *Sex Roles* 12 (1985): 313–328; R. J. Canter and B. C. Meyerowitz, "Sex-Role Stereotypes: Self-Reports of Behavior," *Sex Roles* 10 (1984):293–306.

3. See: Nancy Jay, "Gender and Dichotomy," *Feminist Studies* 7(1981): 37–56. See also: G. Lloyd, *The Man of Reason: "Male"and "Female' "in Western Philosophy* (London: Methuen, 1984), Eleanor Maccoby and Carol Nagy Jacklin, *The Psychology of Sex Differences* (Palo Alto, CA: Stanford University Press, 1974); Marion Lowe, "Social Bodies: The Interaction of Culture and Women's Biology," in *Biological Woman: The Convenient Myth*, eds. R. Hubbard, M. S. Henifin, and B. Fried (Cambridge, MA: Schenkman Publishing Co., 1982), pp. 91–116. See also: C. F. Epstein, *Deceptive Distinctions: Sex, Gender, and the Social Order* (New Haven: Yale University Press, and New York: Russel Sage Foundation, l988), chapter 4.

4. See: Aristotle, "Politicia" and "De Generatione Animalium," in *The Works of Aristotle*, translated by Benjamin Jowelt, eds. W. D. Ross and J. A. Smith (London: Oxford, 1921). Cited in Donna Wilshire, "The Uses of Myth, Image, and the Female Body in Re-Visioning Knowledge," in *Gender/body/Knowledge:Feminist Reconstructions of Being and Knowledge*, ed. Alison M. Jagger and Susan R. Bordo (New Brunswick, NJ: Rutgers University Press,1989), p. 93.

5. Donna Wilshire, "The Uses of Myth, Image, and the Female Body in Re-Visioning Knowledge," in *Gender/Body/Knowledge: Feminist Reconstructions of Being and Knowledge*, eds. Alison M. Jagger and Susan R. Bordo (New Brunswick, NJ: Rutgers University Press, 1989), pp. 92–114.

6. See: Nancy Tuana, "The Weaker Seed: The Sexist Bias of Reproductive Theory," in *Feminism and Science*, ed. Nancy Tuana (Bloomington: Indiana University Press, 1989), pp. 147–171.

7. Thomas Aquinas, "Summa Theologiae,"1:92, in Anthony Synnot. *The Body Social: Symbolism, Self and Society* (New York: Routledge, 1993), p. 46.

8. Cited in Anthony Synnot, *The Body Social: Symbolism, Self and Society* (New York: Routledge, 1993), p. 45.

9. Dawn H. Currie and Valerie Raoul, "The Anatomy of Gender: Dissecting Sexual Difference in the Body of Knowledge," in *Anatomy of Gender: Women's Struggle for the Body*, eds. Dawn H. Currie and Valerie Raoul (Ottawa, Canada: Careleton University Press, 1992), pp. 2–3.

10. Dawn H. Currie and Valerie Raoul, "The Anatomy of Gender: Dissecting Sexual Difference in the Body of Knowledge," in *Anatomy of Gender: Women's Struggle for the Body*, eds. Dawn H. Currie and Valerie Raoul (Ottawa, Canada: Careleton University Press, 1992), pp. 11–34. Currie and Raoul note: "Decartes maintained that the mind exists independently of bodily need and individual experience. He posited that, through the exercise of reason, the thinker could acquire a view of the world which transcends its point of origin. Knowledge achieved through Cartesian reason was thus called objective, in that it is severed from emotional and political considerations, and universal, in that it is able to assume a 'bird's eye' view of the social world" (pp. 1–2).

11. Evelyn Fox Keller offers a psychoanalytic view of how science becomes a male pursuit. Objectivity is not inborn and requires the separation of self from others. The mother plays a crucial role in the child's development of a self. According to Fox-

Keller, "Boys rest their very sexual identity on an opposition to what is both experienced and defined as feminine; the development of their gender identity is likely to accentuate the process of separation." She notes that over time this developmental task results in a firm, somewhat exaggerated sense of autonomy and makes males more suited to the pursuit of objectivity. Girls, on the other hand, develop less of a sense of separation from the mother that may "complicate her development of autonomy by stressing dependency and subjectivity as feminine characteristics." See : Evelyn Fox Keller. "Gender and Science," *Psychoanalysis and Contemporary Thought: A Quarterly of Integrative and Interdisciplinary Studies* 1 (1978): 409–433. See also: Susan Bordo, *The Flight to Objectivity : Essays on Cartesianism and Culture* (Albany: State University of New York Press, 1987). Many of these ideas are theoretically grounded in Nancy Chodorow's work. See Nancy Chodorow, *The Reproduction of Mothering: Psychoanalysis and the Sociology of Gender* (Berkeley: University of California Press, 1978).

12. Francis Bacon, quoted in Carolyn Merchant, *The Death of Nature: Women, Ecology and the Scientific Revolution* (New York: Harper & Row, 1989), p. 168.

13. Zuleyma Tang Halpin, "Scientific Objectivity and the Concept of 'The Other,' " *Women's Studies International Forum* 12 (1989):288.

14. Ruth Berman notes that the dualist rationalism of Aristotle and Plato demonstrates how society's rulers limit and distort the understanding of even profound thinkers in their desire to maintain the status quo in their self-interest. Leaders have historically and currently used the practice of invoking an apparently natural "hierarchy of human worth to justify widely disparate economic and social conditions." See: p. 234, Ruth Berman, "From Aristotle's Dualism to Materialist Dialectics: Feminist Transformation of Science and Society," in *Gender/Body/Knowledge: Feminist Reconstructions of Being and Knowing*, eds. Alison M. Jagger and Susan R. Bordo (New Brunswick, NJ: Rutgers University Press, 1989), pp. 224–255; see also: *Body/Politics: Women and the Discourses of Science*, eds. Mary Jacobus, Evelyn Fox Keller, Sally Shuttleworth (New York: Routledge, 1990).

Many social theorists have remarked on the importance of the body as a central mechanism of social control. See: Michel Foucault, *Discipline and Punish: The Birth of the Prison*, translated by Alan Sheridan (New York: Pantheon Books, 1977); Bryan S. Turner, *The Body and Society* (New York: Basil Blackwell, 1984); *The Body: Social Process and Cultural Theory*, eds. Mike Featherstone, Mike Hepworth, and Bryan S. Turner (Newbury Park, CA: Sage Publications, 1991); Barry Glassner, *Bodies* (New York: Putnam, 1988); Susan R. Bordo, "The Body and the Reproduction of Femininity: A Feminist Appropriation of Foucault," in *Gender/Body/Knowledge: Feminist Reconstructions of Being and Knowing*, eds. Alison M. Jagger and Susan R. Bordo (New Brunswick, NJ: Rutgers University Press 1989), pp. 13–33.

15. See: Susan Lehrer, *Origins of Protective Labor Legislation for Women: 1905–1925* (New York: State University of New York Press, 1987). See also: Zillah R. Eisenstein, *The Female Body and the Law* (Berkeley: University of California Press, 1988).

16. James W. Bashford, *China: An Interpretation* (New York: Abingdon Press, 1916), p. 128. Cited in Susan Greenhalgh, "Bound Feet, Hobbled Lives: Women in Old China," *Frontiers* 2 (1977): 17–21.

17. See: Wolfram Eberhard in "Introduction" in Howard S. Levy, *Chinese Footbinding: The History of a Curious Erotic Custom* (New York: Walton Rawls, 1966); Susan Greenhalgh, "Bound Feet, Hobbled Lives: Women in Old China," *Frontiers* 2 (1977):17–21.

18. Howard S. Levy, *Chinese Footbinding: The History of a Curious Erotic Custom* (New York: Walton Rawls, 1966), pp. 26–27.

19. C. Fred Blake notes that a mother who bound her daughter's feet considered it a sign of caring. He notes: "The 'tradition' could not have passed from mothers to daughters if not for mothers' credibility as 'caring.' The conundrum of a mothers' care consciously causing her daughter excruciating pain is contained in a single word, *teng*, which . . . refers to 'hurting,' 'caring,' or a conflation of both in the same breath" (p. 682). C. Fred Blake," Foot-binding in Neo-Confucian China and the Appropriation of Female Labor," *Signs: Journal of Women in Culture and Society* 19 (1994): 676–712.

20. Fei Hsiao-tung, *China's Gentry: Essays in Rural-Urban Relations* (Chicago: University of Chicago Press, 1953), pp. 32, 84; Chow Yung-ten, *Social Mobility in China: Status Careers Among the Gentry in a Chinese Commmunity* (New York: Atherton Pres, 1966). Both cited in Susan Greenhalgh, "Bound Feet, Hobbled Lives: Women in Old China," *Frontiers* 2 (1977):7–21.

21. Anthropoligist C. Fred Blakes notes, however, that by binding women's feet Chinese society "masked" the real contribution of women's labor to the overall economy: "The material contributions that women made to the family were indeed substantial. They included women's traditional handiwork—making items like clothes and shoes—as well as their biological contributions in making sons for the labor-intensive economy" (p. 700). He notes that binding women's feet, a symbol of their labor power, made it easier for the family system to take over their labor power (pp. 707–708). C. Fred Blake, "Foot-binding in Neo-Confucian China and the Appropriation of Female Labor," *Signs: Journal of Women in Culture and Society* 19 (1994): 676–712.

22. Anthropologist C. Fred Blake notes that " 'big feet' of ordinary women were demeaned as clumsy and crude and as a disaster to the natural foundation—the productivity—of the civilized world" (p. 693). C. Fred Blake, "Foot-binding in Neo-Confucian China and the Appropriation of Female Labor," *Signs : Journal of Women in Culture and Society* 19 (1994): 676–712.

23. Susan Greenhalgh, "Bound Feet, Hobbled Lives: Women in Old China." *Frontiers* 2 (1977):12.

24. See: Andrea Dworkin, *Woman Hating* (New York: Dutton, 1974), pp. 103–104. Anthropologist C. Fred Blake points to several socialization processes that re-inforced the mind/body dualism in Ancient China. He notes: "Boys' and girls' modes of self-realization, of becoming their respective bodies in relationship to others differed completely. The boy's self-realization focused on the locutionary and literary power of the world. The girls' self-realization required her not merely to become, but to 'over-come her body' by restricting the space it filled. . . . The difference was dramatized in innumerable little ways . . . for instance, a daughter having her feet bound might receive a writing brush from her mother. The writing brush was a powerful symbol of masculinity and the world of civil affairs. But unlike her brother, the little girl did not receive the brush with the hope that she might learn how to shape literary discourse. Instead, she grasped the 'point' of the brush in the hope that her feet might acquire its 'pointed' shape" (p. 681). C. Fred Blake, "Foot-binding in Neo-Confucian China and the Appropriation of Female Labor," *Signs : Journal of Women in Culture and Society* 19 (1994): 676–712.

25. Susan Greenhalgh, "Bound Feet, Hobbled Lives: Women in Old China," *Frontiers* 2(1977): 13.

26. See: Susan Greenhalgh, "Bound Feet, Hobbled Lives: Women in Old

China," *Frontiers* 2 (1977):12. See also: Florence Ayscough, *Chinese Women Yesterday and Today* (Boston: Houghton Mifflin Co., 1937), p. 685.

27. Howard S. Levy, *Chinese Footbinding: The History of a Curious Erotic Custom* (New York: Walton Rawls, 1966), p. 19.

28. Howard S. Levy, *Chinese Footbinding: The History of a Curious Erotic Custom* (New York: Walton Rawls, 1966), p. 44.

29. Howard S. Levy, *Chinese Footbinding: The History of a Curious Erotic Custom* (New York: Walton Rawls, 1966), p. 34.

30. Howard S. Levy, *Chinese Footbinding: The History of a Curious Erotic Custom* (New York: Walton Rawls, 1966), p. 32. For an extensive discussion of the erotic nature of the bound foot see: Bernard Rudofsky, *The Kimono Mind* (New York: Doubleday, 1965).

31. Michel Foucault, *Discipline and Punish: The Birth of the Prison*, translated by Alan Sheridan (New York: Pantheon Books, 1977).

32. It is important to point out that males were also subjected to body rituals and practices. Early capitalism with its mechanization of production needed disciplined bodies to do the mundane work routines in the early factory system. Using the body as a central arena of disciplinary power and control allowed nineteenth century capitalism to operate efficiently and profitably. As Dreyfus and Rabinow note, "Without the insertion of disciplined, orderly individuals into the machinery of production, the new demands of capitalism would have been stymied." Hubert L. Dreyfus and P. Rabinow, *Michel Foucault: Beyond Structuralism and Hermeneutics* (Chicago: The University of Chicago Press, 1983), p. 135. A workers' body (women workers were considered a lower-paid category compared with males) was equated with that of a machine. In his approach, known as "Taylorism," Frederick Taylor, founder of "scientfic management," envisioned a worker as part of the machinery of production. Through the application of scientific principles, specifically the technique of "time and motion studies," he proposed to ascertain how to get the most efficiency out of a given worker, ignoring some of the humanistic aspects of work. See: Frederick Taylor, *Principles of Scientific Management* (New York: W.W. Norton & Co., 1967).

33. Ann Gordon, Mari Jo Buhle, and Nancy Schrom, "Women in American Society: An Historical Contribution," *Radical America* 5 (July–August 1971): 3–66.

34. Kathryn Weibel, *Mirror, Mirror: Images of Women Reflected in Popular Culture* (New York: Anchor Books, 1977), pp.176–177. Kathryn Weibel argues that the separation in roles fostered by the Industrial Revolution was reflected in the disparity in comfort and ornamentation between men's and women's clothes. The Industrial Revolution generated a larger, relatively wealthy, middle class of men. As has been the case historically, wives were expected to display the wealth of their husbands, becoming more "ornamented" and more "stuffed-looking" as middle-class wealth increased during the nineteenth century.

35. Fashion historian Valerie Steele notes that "the vast majority of women of all classes wore corsets and the degree of tightness varied according to design of dress, social occasion and age, personality and figure of the individual woman." See: Valerie Steele, *Fashion and Eroticism: Ideals of Feminine Beauty from the Victorian Era to the Jazz Age* (New York: Oxford University Press, 1985), p. 162. See also: Helene E. Roberts, "The Exquisite Slave: The Role of Clothes in the Making of the Victorian Woman," *Signs: Journal of Women in Culture and Society* 2 (1977): 554–569.

36. Kathryn Weibel, *Mirror, Mirror: Images of Women Reflected in Popular Culture* (New York: Anchor Books, 1977), p. 180. While establishing links between social or

cultural influences and illness is difficult, it has been suggested that a form of anemia, known as chlorosis, reflected the cultural repression women experienced during the Victorian Era. Joan Brumberg notes that chlorosis, an illness characterized by weakness, fainting, and passivity was widespread among young women in the United States dating from 1870 to 1920. See: Joan J. Brumberg, "Chlorotic Girls, 1870–1920: A Historical Perspective on Female Adolescence," *Child Development* 53 (1982):1468–1477. Other researchers (Donald M.Schwartz, Michael G. Thompson, and Craig L. Johnson, "Anorexia Nervosa and Bulimia: The Socio-Cultural Context," *International Journal of Eating Disorders* 1 (1982): 20–36) suggest that rates of classical conversion hysteria may also be another example of the importance of cultural pressures—in this case an environment where sexual repression and dependency were primary characterizations of women's role.

37 Mlle. Pauline Mariette, *L'Art de la toilette* (Paris: Librairie Centrale, 1866), pp. 40–4l. Cited in Valerie Steele, *Fashion and Eroticism: Ideals of Feminine Beauty from the Victorian Era to the Jazz Age* (New York: Oxford University Press, 1985), p. 108.

38. Valerie Steele, *Fashion and Eroticism: Ideals of Feminine Beauty from the Victorian Era to the Jazz Age* (New York: Oxford University Press, 1985), p. 114.

39. Lawrence Stone compares corset makers with those who practice orthodontia. He notes that corset makers were "the affluent equivalents of the orthodontists of the late twentieth-century America, who also cater for a real need as well as a desire for perfection in a certain area thought to be important for success in life." See Lawrence Stone, *The Family, Sex and Marriage in England 1500–1800* (New York, Harper & Row, 1977), as cited in William Bennett and Joel Gurin, *The Dieter's Dilemma: Eating Less and Weighing More* (New York: Basic Books, 1982), p. 183. Another researcher points out that many medical professionals were not very aware and concerned with the effects of tight lacing. One historian hypothesizes that women who fainted were not necessarily suffering from psychosomatic illnesses, but the effects of tight lacing. See: Mel Davies, "Corsets and Conception: Fashion and Demographic Trends in the Nineteenth Century," *Comparative Studies in Sociology and History* 24 (1982): 611–641.

40. *Englishwoman's Domestic Magazine*, 3d ser. 4 (1868): 54. Cited in Helene E. Roberts, "The Exquisite Slave: The Role of Clothes in the Making of the Victorian Woman," *Signs: Journal of Women in Culture and Society* 2 (1977):564.

41. See: Lorna Duffin, "The Conspicuous Consumptive: Woman as an Invalid," in *The Nineteenth Century Woman, Her Cultural and Physical World*, eds. Sara Delamont and Lorna Duffin (London: Croom Helm, 1978), pp. 26–56; Barbara Ehrenreich and Deirdre English, *For Her Own Good: 150 Years of the Experts' Advice to Women* (Garden City, NY: Anchor Books, 1979); Helene E. Roberts, "The Exquisite Slave: The Role of Clothes in the Making of the Victorian Woman," *Signs: Journal of Women in Culture and Society* 2 (1977): 554–569; Thorstein Veblen, *The Theory of the Leisure Class* (New York: Random House, The Modern Library, original work published 1899).

There are caveats on interpreting the corset as primarily a means for subordinating women. Valerie Steele notes: "The idea that nineteenth century (male dominated) society forced women into submissive and masochistic behavior is not really substantiated by the sartorial and documentary evidence. Even though most woman were economically dependent on men, and may have needed to conform to male ideals to a certain extent, women's self-images and sexuality were not completely male defined. . . . My own research has indicated that the clothing of the Victorian woman reflected not only the cultural prescriptive ideal of femininity but also her own aspirations and fantasies. . . . . The Victorian woman played many often contradictory and ambiguous roles, but she cannot be characterized as a prude, a masochist or a slave." Valerie Steele, *Fashion and Eroticism: Ideals of Feminine Beauty from the Victorian Era to the Jazz Age* (New York: Oxford University Press, 1985), pp. 100–101.

While many historians saw the corset as a means of socially controlling women, not all are in agreement with this thesis. One historian argues that those who opposed corseting in the nineteenth century were socially conservative males who felt tight-lacing was a symbol of women's resistance to male authority (p. 579). From this perspective, tight-lacing provided women with a means to express their sexuality and to gain some degree of power over their oppressive situation. David Kunzel notes: "In vain the preachers threatened women who exposed their breasts with cancer of the breast. In vain the preachers and physicians threatened provocatively corseted women with every anathema, disease, and even death itself" (p. 574). See: David Kunzle, "Dress Reform as Antifeminism: A Response to Helene E. Robert's 'The Exquisite Slave: The Role of Clothes in the Making of the Victorian Woman,' " *Signs* 2,3 (1977): 570–579. See also: Valerie Steele, *Fashion and Eroticism: Ideals of Feminine Beauty from the Victorian Era to the Jazz Age* (New York: Oxford University Press, 1985), pp. 161–191.

42. Kathryn Weibel, *Mirror, Mirror: Images of Women Reflected in Popular Culture* (New York: Anchor Books, 1977), p. 179.

43. Helene E. Roberts, "The Exquisite Slave: The Role of Clothes in the Making of the Victorian Woman," *Signs* 2 (1977): 557. Medical practitioners became primary controllers of women's bodies during the Victorian era. For example, women who showed signs of excessive "sexual excitement" (usually considered a problem of middle and upper middle-class women) were prime candidates for control. Other symptoms subject to medical control included those women who exhibited signs of "troublesomeness," those who were "eating like a ploughman," masturbated, attempted suicide, who suffered from "persecution mania," simple "cussedness" and dysmenorrhea. The cure often involved several of the following techniques: removal of the ovaries and/or clitoris, hot steel applied to the cervix, and/or leeches placed on the womb. See Barbara Ehrenreich and Deidre English, *For Her Own Good: 150 Years of the Experts' Advice to Women* (Garden City, NY: Anchor Books, 1979), p. 124.

44. Helene E. Roberts, "The Exquisite Slave: The Role of Clothes in the Making of the Victorian Woman," *Signs: Journal of Women in Culture and Society* 2 (1977): 564.

45. William Bennett and Joel Gurin, *The Dieter's Dilemma: Eating Less and Weighing More* (New York: Basic Books, 1982), p. 183.

46. Lawrence Stone, *The Family, Sex and Marriage in England 1500–1800* (New York: Harper & Row, 1977), as cited in William Bennett and Joel Gurin, *The Dieter's Dilemma: Eating Less and Weighing More* (New York: Basic Books, 1982), p. 183.

47. It is important to note that patriarchal interests were also threatened during early industrialism. Capitalism challenged patriarchal power by separating the home from the workplace. Barbara Ehrenreich and Deidre English note: "The household was left with only the most biological activities—eating, sex, sleeping, the care of small children . . . birth, dying and the care of the sick and aged." Furthermore, "It was now possible for a woman to enter the market herself and exchange her labor for the means of survival." See: Barbara Ehrenreich and Deidre English, *For Her Own Good:*

150 Years of the Experts' Advice to Women (Garden City, NY: Anchor Books, 1979): pp. 10, 13, 27.

48. Mike Featherstone, "The Body in Consumer Culture," *Theory, Culture and Society,* 2:(1982):20. See also: Stuart Ewen, *Captains of Consciousness: Advertising and the Roots of the Consumer Culture* (New York: McGraw-Hill, 1976); Joseph Hansen and Evelyn Reed, *Cosmetics, Fashions and the Exploitation of Women* (New York: Pathfinder Press, 1986); Heidi Hartmann, "Capitalism, Patriarchy and Job Segregation by Sex," *Signs: Journal of Women in Culture and Society* 1 (1976): 137–169.

49. Roberta Seid provides an excellent detailed historical analysis of American society's movement toward slenderness. See especially Chapter 5, Roberta Pollack Seid, *Never Too Thin: Why Women Are at War with Their Bodies* (New York: Prentice-Hall Press, 1989). See also: Stuart Ewen and Elizabeth Ewen, *Channels of Desire: Mass Images and the Shaping of the American Consciousness* (New York: McGraw-Hill, 1982).

50. Roberta Pollack Seid, *Never Too Thin: Why Women Are at War with Their Bodies* (New York: Prentice Hall, 1989), p.85.

51. Roberta Pollack Seid, *Never Too Thin: Why Women Are at War with Their Bodies* (New York: Prentice Hall, 1989), p. 83.

52. Roberta Pollack Seid, *Never Too Thin: Why Women Are at War with Their Bodies* (New York: Prentice Hall, 1989), p. 115.

53. Roberta Pollack Seid, *Never Too Thin: Why Women Are at War with Their Bodies* (New York: Prentice Hall, 1989), p. 115.

54. Fred Davis, *Fashion, Culture, and Identity* (Chicago: The University of Chicago Press, 1992), p. 39.

55. Barbara Ehrenreich and Deidre English, *For Her Own Good: 150 Years of the Experts Advice to Women* (Garden City, NY: Anchor Books, 1979); Stuart Ewen, *Captains of Consciousness: Advertising and the Roots of Consumer Culture* (New York: McGraw-Hill, 1976); Joseph Hansen and Evelyn Reed. Cosmetics, *Fashions and the Exploitation of Women* (New York: Pathfinder Press, 1986) ; Heidi Hartmann, "Capitalism, Patriarchy, and Job Segregation by Sex," *Signs: Journal of Women in Culture and Society* 1 (1976): 137–169; Brett Silverstein, *Fed Up! The Food Forces That Make You Fat, Sick and Poor* (Boston: South End Press, 1984).

56. Sharlene Hesse-Biber, "Women, Weight and Eating Disorders: A Socio-Cultural and Political-Economic Analysis," *Women's Studies International Forum* 14 (1991): 173–191; Susan Bordo, *Unbearable Weight: Feminism, Western Culture and the Body* (Berkeley: University of California Press, 1993).

57. Mary Frank Fox and Sharlene Hesse-Biber, *Women at Work* (Palo Alto, CA.: Mayfield Publishing Co., 1984), p.19.

58. In the late 1800s, a large number of women entered medicine. Of the 1893–94 medical graduating classes in the Boston area 23.7% were women and 17% of Boston's medical community were women. But, but the end of World War I, the numbers of women dropped off. See: Augusta Greenblatt, "Women in Medicine," *National Forum. The Phi Beta Kappa Journal* 61 (1981): 10–11.

59. For a detailed discussion of this transition, see: Lois Banner, *American Beauty.*(Chicago: The University of Chicago Press, 1983) and William Bennett and Joel Gurin, *The Dieter's Dilemma: Eating Less and Weighing More* (New York: Basic Books,Inc.,1982), Chapter seven, "The Century of Svelte."

60. See: Wendy Chapkis, *Beauty Secrets: Women and the Politics of Appearance* (Boston: South End Press, 1986), pp. 15–16.

61. See: Susie Orbach, *Hunger Strike: The Anorectic's Struggle as a Metaphor of Our Age* (New York: W.W. Norton, 1986), p. 75.

62. Roberta Pollack Seid, *Never Too Thin: Why Women are at War With Their Bodies* (New York: Prentice-Hall, 1989).

63. See Chapter 5, Roberta Pollack Seid, *Never Too Thin: Why Women Are at War with Their Bodies* (New York: Prentice-Hall, 1989).

64. See: Lois W. Banner, *American Beauty* (New York: Knopf, 1983), p. 279.

65. John A. Ryle, "Discussion of Anorexia Nervosa," *Proceedings of the Royal Society of Medicine* 32 (1939): 735–737. It is important to note that documented cases of anorexia nervosa were cited in the medical literature well before this time. See William Gull, "Anorexia Nervosa (Apepsia Hysterica, Anorexia Hysterica)," *Transactions of the Clinical Society of London* 7 (1974): 22–28, and (English language translation): Ernest-Charles Lasègue, "On Hysterical Anorexia," *Medical Times and Gazette* (6 Sept. 1873): 265–266; (27 Sept.): 367–369. Historian Edward Shorter points out, however, "It is only in the decade before the first world war that references to anorexia in aid of modish thinness and romantic acceptance begin to proliferate." Edward Shorter, "The First Great Increase in Anorexia Nervosa," *Journal of Social History* 21 (1987): 82. Roberta Pollack Seid notes that even though there was a "slenderness craze" during this time (which she dates from 1919 to 1935), the pursuit of thinness was very different from that of the post 1960s decades: "the craze did not create the hysteria or the terrible effects on women that we know today. The positive associations with plumpness were too recent and too well entrenched to be easily eradicated." See: Roberta Pollack Seid, *Never Too Thin: Why Women Are at War with their Bodies* (New York: Prentice Hall, 1989) p. 97.

66. "Weight Reduction Linked to the Mind," *New York Times* (February 24, 1926).

67. Richard A. Gordon, *Anorexia and Bulimia: Anatomy of a Social Epidemic* (Cambridge, MA: Basil Blackwell, 1990), p. 78.

68. See: William Bennett and Joel Gurin, *The Dieter's Dilemma: Eating Less and Weighing More* (New York: Basic Books, Inc., 1982), p. 207.

69. Lois W. Banner, *American Beauty* (New York: Knopf, 1983), p. 283.

70. This is amply documented in Betty Friedan, *The Feminine Mystique* (New York: W.W. Norton, 1963).

71. Historian Lois Banner notes that during the 1950s women's sports suffered a setback. "With few exceptions, the kind of acclaim accorded to individual women sports stars in the 1920s and 1930s no longer existed, and the commercial women's swimming and basketball teams popular in these earlier decades faded from view. . . . In high schools and colleges, women's athletics similarly came to occupy a modest position vis-à-vis men's sports. See: Lois W. Banner, *American Beauty* (New York: Knopf, 1983), p. 285. During the 1950s there was evidence of the flapper in the image of Debbie Reynolds and Sandra Dee. Sandra Dee played the popular Gidget who was portrayed as looking for a husband and not serious about a career. See: Lois W. Banner, *American Beauty* (New York: Knopf, 1983), p. 283.

72. Roberta Pollack Seid, *Never Too Thin: Why Women Are at War with Their Bodies* (New York: Prentice-Hall, 1989), p. 257.

73. Mary Frank Fox and Sharlene Hesse-Biber, *Women at Work* (Palo Alto, CA: Mayfield Publishing Co., 1984).

74. An empirical test of this theory on changing body image comes from a study by Silverstein, Perdue, Peterson, Vogel, and Fantini (1986). They studied the stan-

dards of bodily attractiveness across time and note that over the course of the twentieth century, as the proportion of American women who worked in the professions or
who graduated from college increased, the standard of bodily attractiveness became
less curvaceous. They note that this occurred especially in the 1920s and during the
1960s. Thinness may be considered a sign of conforming to a constricting feminine
image (like corseting), whereas greater weight may convey a strong, powerful image.
See: Brett Silverstein, Lauren Perdue, Barbara Peterson, Linda Vogel, and Deborah A.
Fantini. "Possible Causes of the Thin Standard of Bodily Attractiveness for Women,"
International Journal of Eating Disorders 5 (1986), 135–144.

75. Lois W. Banner, American Beauty (New York: Knopf, 1983), pp. 266–287.

76. Allan Mazur, "U.S. Trends in Feminine Beauty and Overadaptation," Journal
of Sex Research 22 (1986): 281–303.

77. Doug Stewart, "In the Cutthroat World of Toy Sales, Child's Play Is Serious
Business," Smithsonian 20 (December, 1989): 80.

78. Doug Stewart, "In the Cutthroat World of Toy Sales, Child's Play Is Serious
Business," Smithsonian 20 (December, 1989): 72–84.

79. Marjorie Ferguson, Forever Feminine: Women's Magazines and the Cult of
Femininity (London: Heinemann, 1983), p. 184.

80. Susie Orbach, Fat Is a Feminist Issue (New York: Berkeley Press, 1978), p. 21.

81. See: Kim Chernin, The Obsession: Reflections on The Tyranny of Slenderness
(New York: Harper & Row, 1981), p. 110.

82. Pauline B. Bart, "Emotional and Social Status of the Older Woman," in No
Longer Young: The Older Woman in America. Proceedings of the 26th Annual Conference
on Aging. eds. Pauline Bart et al. (Ann Arbor: University of Michigan, Institute of
Gerontology, 1975), p. 321; Daniel Bar-Tal, and Leonard Saxe, "Physical Attractive-
ness and Its Relationship to Sex-Role Stereotyping," Sex Roles 2 (1976): 123–133;
Peter Blumstein and Pepper W. Schwartz, American Couples: Money, Work and Sex
(New York William Morrow, 1983); Glen H. Elder, "Appearance and Education in
Marriage Mobility," American Sociological Review 34 (1969): 519–533; Susan Sontag,
"The Double Standard of Aging," Saturday Review 55 (1972): 29–38.

83. Sharlene Hesse-Biber, Alan Clayton-Matthews, and John Downey, "The
Differential Importance of Weight Among College Men and Women," Genetic, Social
and General Psychology Monographs 113 (1987): 511–528.

84. David Kunzle, "Dress Reform as Antifeminism: A Response to Helene E.
Roberts' 'The Exquisite Slave: The Role of Clothes in the Making of the Victorian
Woman," Signs: 2 (1977): 570–579. The wearing of men's clothing in the 1930s and
1940s and the "Annie Hall" look in the late 1970s and early 1980s might be inter-
preted as a way women resisted the dominant fashion trends that sought to control
them. One researcher notes: "The wearing of men's clothes can mean many different
things. In the thirties, sophisticated actresses such as Marlene Dietrich in top hat and
tails and elegantly cut suits projected sophistication, power and a dangerous eroticism.
The slacks and sweaters of the war period, and the jeans and pants outfits of the sixties
and early seventies, were serious gestures toward sexual equality (p. 229). "However,
these images, especially the Annie Hall look, could turn into an "ironic antifeminest
message," as this researcher also notes. "Because they are worn several sizes too large,
they suggest a child dressed up in her daddy's or older brother's things for fun, and
imply 'I'm only playing; I'm not really big enough to wear a man's pants, or do a man's
job.' " See: Alison Lurie, The Language of Clothes (New York: Vintage Books, 1983),
p. 229.

85. Barbara Ehrenreich and Deirde English, *For Her Own Good: 150 Years of the Experts' Advice to Women* (Garden City, NY: Anchor Books, 1979); Zillah R. Eisenstein, *The Female Body and the Law* (Berkeley: The University of California Press,1988); Emily Martin, *The Woman in the Body: A Cultural Analysis of Reproduction* (Boston: Beacon Press, 1987); Helena Michie, *The Flesh Made Word: Female Figures and Women's Bodies* (New York: Oxford University Press. 1987); Gayle Rubin, "The Traffic in Women," in *Toward an Anthropology of Women*, ed. Rayna R. Reiter (New York: Monthly Review Press, 1975), pp. 157–210.; Bryan S. Turner, *The Body and Society* (New York: Basil Blackwell, Inc., 1984); Susan Bordo, "Anorexia Nervosa: Psychopathology as the Crystallization of Culture," in *Feminism and Foucault: Reflections on Resistance*, eds. Irene Diamond and Lee Quinby (Boston, MA: Northeastern University Press, 1988), pp. 87–117.

86. Michel Foucault, *Discipline and Punish: The Birth of Prison*, translated by Alan Sheridan (New York: Pantheon Books, 1977).

Chapter 3

1. *The American Heritage Dictionary of the English Language* (New York: American Heritage Publishing Co, Inc., and Houghton Mifflin Company, 1973), p. 837.

2. Ilana Attie and J. Brooks-Gunn, "Weight Concerns as Chronic Stressors in Women," in *Gender and Stress*, eds. Rosalind K. Barnett, Lois Biener, and Grace Baruch (New York: The Free Press, 1987), pp. 218–252.

3. Katha Pollitt, "The Politically Correct Body,"*Mother Jones* (May 1982):67. I don't want to disparage the positive benefits of exercising and the positive self-image that can come from feeling good about one's body. This positive image can spill over into other areas of one's life, enhancing, for example, one's self-esteem, or job prospects.

4. See Lawrence D. Cohn and Nancy E. Adler, "Female and Male Perceptions of Ideal Body Shapes: Distorted Views Among Caucasian College Students," *Psychology of Women Quarterly* 16 (1992): 69–79. A. Fallon and P. Rozin, "Sex Differences in Perceptions of Desirable Body Shape," *Journal of Abnormal Psychology*, 94 (1985): 102–105.

5. *Vogue*, "How to Look Like a Beauty" (September 15, 1957): 156.

6. *Cosmopolitan* (December, 1992): 12.

7. Cited in: B. F. Liebman, "Fated to be Fat?," *Nutrition Action Health Letter* 14 (January/February, 1987):4.

8. Brett Silverstein, *Fed Up!* (Boston: South End Press, 1984), pp. 4, 47, 110. Individuals may be affected in many different ways, from paying too much (in 1978, concentration within the industry led to the overcharging of consumers by $12 to $14 billion [p. 47]) to the ingestion of unhealthy substances.

9. Hillel Schwartz, *Never Satisfied* (New York: Free Press, 1986), p. 245.

10. Hillel Schwartz, *Never Satisfied* (New York: Free Press, 1986), p. 264.

11. Hillel Schwartz, *Never Satisfied* (New York: Free Press, 1986), p. 245.

12. J. Dagnoli and J. Liesse, "Kraft, ConAgra Go Head-to-Head in Healthy Meals," *Advertising Age* (October 22, 1990): 59.

13. Warren J. Belasco, " 'Lite' Economics: Less Food, More Profit," *Radical History Review* 28–30 (1984):254–278. Hillel Schwartz, *Never Satisfied* (New York: Free Press, 1986), p. 241.

14. Warren J. Belasco, " 'Lite' Economics: Less Food, More Profit,"*Radical History Review* 28–30 (1984):270.

15. Julie Liesse, "Healthy Choice Growing Pains: Why ConAgra Will Spend $200M on Energizing Its Megabrand," *Advertising Age* 63 (August 24, 1992):3.

16. Kathleen Deveny, " 'Light' Foods Are Having Heavy Going," *Wall Street Journal* (March 4, 1993): B1.

17. George Lazarus, "Nestle Thaws Out Hearty Portions to Beef Up Its Frozen-Entree Menus," *Chicago Tribune* (May 11, 1995):2.

18. Gabriella Stern, "Makers of Frozen Diet Entrees Start Some Diets of Their Own." *Wall Street Journal* (January 4, 1994): B10

19. Chip Walker, "Fat and Happy," *American Demographics* (January 1993): 53–54.

20. Data from this study cited in a *Boston Globe* article. Alison Bass, "Record Obesity Levels Found," *Boston Globe*. (July 20, l994): l, 10.

21. Russell Mitchell, Lois Therrien, and Gregory L. Miles, "ConAgra: Out of the Freezer," *Business Week* (June 25, 1990):24.

22. J. Dagnoli and J. Liesse, "Kraft, ConAgra Go Head-to-Head in Healthy Meals," *Advertising Age* (October 22, 1990): 59.

23. Kraftco, long the maker of various dairy products (including Kraft Macaroni and Cheese, Velveeta, Sealtest ice cream, creamy salad dressings, and the like), decided in 1989 to produce a no-fat version of many of its products, including Entenmann's desserts, cheese slices, dressings, and ice cream. Kraft's profit jumped 26% to 2.1 billion dollars that year, partly due to its marketing of no-fat products. (p.l00) in Lois Therrien, "Kraft Is Looking for Fat Growth from Fat-Free Foods," *Business Week*. (March 26, 1990): 100–101. However, having no-fat versions of lots-of-fat products was only the beginning. Kraft also created Eating Right and Budget Gourmet Light and Healthy frozen dinners. See: J. Dagnoli and J. Liesse, "Kraft, ConAgra Go Head-to-Head in Healthy Meals," *Advertising Age* (October 22, 1990):59. Since Kraft was purchased by cigarette and (through Miller) beer giant Phillip Morris, the same people who profit from Eating Right and Entenmann's no-fat chocolate cake also profit from Jello, Kool-Aid, and the sugar-coated cereals that General Foods (now owned by Phillip Morris) creates.

24. Brett Silverstein, *Fed Up* (Boston: South End Press, 1984), p. 13.

25. Jim Hightower, *Eat Your Heart Out: Food Profiteering in America* (New York: Crown Publishers, Inc., 1975), p. 9.

26. Lois Therrien, "The Food Companies Haven't Finished Eating," *Business Week* (January 9, 1989):70; Lois Therrien explains further: "In the past three years, for example, Borden Inc. has increased its share of the pasta market from 10 percent to 31 percent, largely by snapping up regional producers." Therrien further notes that in 1989, the ten biggest food processors accounted for 35% of all U.S. food shipments, and the largest 20 companies widened their gross profits from 27% in 1982 to 35% in 1988. Lois Therrien, "The Food Companies Haven't Finished Eating," *Business Week* (January 9, 1989):70.

27. Jeffrey Schrank, *Snap, Crackle and Popular Taste* (New York: Dell, 1977), p. 48.

28. Edward Giltenan, "Food, Drink, Tobacco," *Forbes* (January 8, 1990): 172–174. However, in order to more fully understand both the workings of the food industry and the ironies of diet food, it can be helpful to consider specific companies. Beatrice Foods started acquiring companies in 1894, and by 1984 had acquired

more than 400 [Brett Silverstein, *Fed Up!* (Boston: South End Press, 1984), p. 5]. Some Beatrice companies have included Wesson Oil, Butterball Turkey, Hunt's Tomato Products, Tropicana Fruit Drinks, Peter Pan Peanut Butter, Orville Reddenbacher popcorn, La Choy Chinese food, and Rosarita Mexican Products [Lois Terrien, "Beatrice Investors Will Just Have to Sit Tight," *Business Week* (March, 12, 1990):104]. ConAgra, owner of Morton Products and Chun King among many others, had a net income of over 200 million dollars in 1989 [Edward Giltenan, "Food, Drink and Tobacco," *Forbes* (January 8, 1990):172]. On June 7, 1990, ConAgra purchased Beatrice Foods for 2.34 billion dollars [Russell Mitchell, Lois Therriren, and Gregory Miles, "ConAgra: Out of the Freezer," *Business Week* (June 25, 1990): 24]; with the purchase of Beatrice, ConAgra became a 20-billion-dollar corporation [Russell Mitchell, Lois Therrien, and Gregory Miles, "ConAgra: Out of the Freezer," *Business Week.* (June 25, 1990): 25]. Most people are probably not aware of this buyout, nor of the range of products that each company made and makes (*Moody's Industrial Manual 1993*, Vol. 2, Moody's Investor Services, New York, 1993). p.3399.

Sara Lee, with 11.7 billion dollars in sales in 1988 [Robert McGough, "Icing on the Cake," *Financial World.* (October 17, 1989): 22], had a net income of over 400 million dollars in 1989 [Edward Giltenan, "Food, Drink and Tobacco," *Forbes* (January 8, 1990):172]. Unknown to most people, Sara Lee controls 42% of the hosiery market through its Hanes and L'Eggs companies and owns Hillshire Farm and Jimmy Dean sausages, Isotoner gloves, Hygrade hot dogs, Bali bras, Kiwi shoe polish, and PYA/Monarch food service, and even sells Fuller Brush products through the mail. We think of Sara Lee as a bakery products company, but it is second in the packaged meat segment, just behind Oscar Mayer (which was once owned by General Foods and is now owned by Philip Morris) Robert McGough, "Icing on the Cake," *Financial World* (October 17, 1989): 22–24.

HJ Heinz, with over 1,500 products, was worth 6.3 billion dollars in 1989, and accounted for more than 50% of the U.S. ketchup market at that time. Heinz also owns Ore-Ida frozen potatoes, which provides half of the frozen potatoes in the U.S. retail potato market, and Star-Kist, responsible for 37% of the tuna sector [A. Gabb, "Heinz Meanz Brandz," *Management Today* (July, 1989): 64–66]. Heinz had a net income of more than 400 million dollars in 1989 [Edward Giltenan, "Food, Drink and Tobacco," *Forbes* (January 8, 1990): 172]. Heinz also owns Weight Watchers.

29. Lois Therrien, "The Food Companies Haven't Finished Eating," *Business Week* (January 9, 1989):70.

30. "Losing Weight: A Profitable Business," *Christian Science Monitor* (October 8, 1992), p.8.

31. Advertised in *Parade* Magazine (December 30, 1984).

32. John La Rosa is the research director of Market Data Enterprises, a market research company in Valley Stream, L.I., specializing in the diet industry. Cited in: Elizabeth Rosenthal, "Commercial Diets Lack Proof of their Long-Term Success," *New York Times* (November 24, 1992): p.1, p. C11.

33. Deralee Scanlon, *Diets That Work* (Chicago: Contemporary Books, 1991), p. 1.

34. Jenny Craig, The Diet Center, and Nutri/System use similar strategies for programs that range from nine weeks to over seventeen weeks, and include the purchase of special foods to support the weight-loss program. As an example, in 1989, Jenny Craig cost $1,000 to $1,225 for an average seventeen-week diet program in

which an individual would lose one and one-half to two and one half pounds per week (p. 57). See: Annetta Miller, Karen Springen, Linda Buckley, and Elisa Williams, "Diets Incorporated," *Newsweek* (September 11, 1989):56–60.

35. C. Sanz and L. F. Mitchell," Fitness Tycoon Jenny Craig Turns Weight Losses into Profit by Shaping her Clients' Bottomline," *People* (February 19, 1990): 91–92.

36. Brian O'Reilly, "Diet Centers are Really in Fat City," *Fortune* (June 5, 1989): 137.

37. The Diet Workshop is similar to Weight Watchers, but has smaller groups. During the summer of 1995, the first 4 weeks of meetings cost $63. In general, the registration visit is $15, and all following visits are $12. There is a special program called Person to Person, which involves more one-on-one nutritional counseling and costs $99 for four weeks [personal communication]. As of 1990, there were 1,500 Diet Workshop centers in the U.S. [Theodore Berland, "Rating the Weight Loss Clinics," *Consumers Digest* (May/June, 1990): 67]. The Diet Workshop boasts the same reduction rate as Weight Watchers and is currently slightly cheaper.

TOPS, Take Off Pounds Sensibly, relies on intragroup competition and camaraderie, and focuses less on nutritional education than Weight Watchers or the Diet Workshop. There are 11,873 chapters worldwide [Theodore Berland," Rating the Weight Loss Clinics," *Consumers Digest* (May/June 1990): 67], with more than 18,000 members in California alone [Deralee Scanlon, *Diets That Work* (Chicago: Contemporary Books, 1991), 20]. TOPS is cheaper than the other two mentioned, with $12–$16 in annual fees, and weekly dues averaging fifty cents a meeting [Matthew Quincy, *Diet Right!* (Berkeley, CA: Conari Press, 1991), p. 121; Deralee Scanlon, *Diets That Work* (Chicago: Contemporary Books, 1991), p. 21].

38. Weight Watchers International, Inc., Corporate Backgrounder, 1994.

39. Matthew Quincy, *Diet Right!* (Berkeley, CA: Conari Press, 1991), p. 55.

40. Weight Watchers International, Inc., Corporate Backgrounder, 1994.

41. Weight Watchers International, Inc., Corporate Backgrounder, 1994.

42. Brian O'Reilly, "Diet Centers Are Really in Fat City," *Fortune* (June 5, 1989): 140.

43. Annabella Gabb, "Heinz Means Brandz," *Management Today* (July 1989): 68.

44. Annetta Miller et al., "Diets Incorporated," *Newsweek* (September 11, 1989): 59.

45. Sandy Lutz, "Weight Loss Market's Profits Are Fading," *Modern Healthcare* (February 19, 1990): 50.

46. Annetta Miller et al., "Diets Incorporated," *Newsweek* (September 11, 1989): 60.

47. Sandy Lutz, "Weight Loss Market's Profits Are Fading," *Modern Healthcare* (February 19, 1990): 50.

48. Annetta Miller et al., "Diets Incorporated," *Newsweek* (September 11, 1989): 57. One of the most popular programs, Slim Fast, is actually a "combo" program, involving two liquid meals a day and one nutritionally balanced regular meal. Slim Fast users doubled in 1988 to 20 million. The company's 1994 revenues were estimated at $650 million (Gordon Verrell, "Lasorda's Loss," *The Sporting News* [March 6, 1995]:7). Slim-Fast has now begun a magazine similar to the one put out by Weight Watchers, and its diet frozen dinners are beginning to appear in supermarkets.

There are, of course, people whom these programs have not helped, and such people might go to further lengths to lose weight. They might go to the Duke

University Diet and Fitness Center, which costs $7,000 for 7 weeks [Thomas DeFrank, "Tales from the Diet Trenches," *Newsweek* (September 11, 1989)] or to Feeding Ourselves, a Boston-based program for compulsive eaters, which offers a summer weekend intensive session for $225. The Pritikin Longevity Center, located in Santa Monica, California, and Miami Beach, Florida, offered 7, 13, and 26 day residential programs for $3,200, $5,500, and $9,000 in 1989, with an advertised weight loss of 2 to 4 pounds a week [Annetta Miller et al., "Diets Incorporated," *Newsweek* (September 11, 1989): 57]; Canyon Ranch, a spa in Tucson, Arizona, and Lenox, Massachusetts, offered in December 1994 a 3 night/4 day program costing $984 for a double room and $1,190 for a single suite.

49. William Bennett and Joel Gurin, *The Dieter's Dilemma: Eating Less and Weighing More* (New York: Basic Books, Inc., 1982), p. 238.

50. Elisabeth Rosenthal, "Commercial Diets Lack Proof of Their Long Term Success," *New York Times* (November 24, 1992): Al, C11.

51. "Hypothalamic Set-Point System May Regulate Weight Loss," *American Family Physician* (March 1984): 269.

52. See: Stanton Peele, *Diseasing of America: Addiction Treatment Out of Control* (Lexington, MA: D.C. Heath and Co., 1989).

53. There are a few recovery books that point to the larger issues of the addiction model. Anne Wilson Schaef's book, *When Society Becomes an Addict*, looks at the wider institutions of society that perpetuate addiction. She notes that society operates on a scarcity model. This is the "Addictive System." This model assumes that there is never enough of anything to go around and we need to get what we can. Schaef sees society as made up of three systems: A White Male System (the Addictive System), A Reactive Female System (one where women respond passively to men by being subject to their will), and the Emerging Female System (a system where women lead with caring and sensitity). Society needs to move in the direction of the Emerging Female System in order to end addiction. Another important book is Stanton Peele's *Love and Addiction*. Another book by Stanton Peele, *The Diseasing of America: How the Addiction Industry Captured Our Soul* (Lexington, MA: Lexington Books, 1989), stresses the importance of social change in societal institutions and advocates changing the given distribution of resources and power within the society as a way to overcome the problem of addiction. See: Anne Wilson Schaef, *When Society Becomes an Addict* (New York: Harper & Row, 1987), and Stanton Peele, *Love and Addiction* (New York: New American Library, 1975).

54. Neal Karlen, "Greetings from MINNESOBER!," *New York Times* (May 28, 1995):32; Margaret Jones, "The Rage for Recovery," *Publisher's Weekly* (November 23, 1990):16–24.

55. Wendy Kaminer, "Chances Are You're Co-Dependent Too," *New York Times Book Review* (February 11, 1990): 26.

56. Gayle Feldman, "On the Road to Recovery with Prentice Hall, Ballantine, et al," *Publishers Weekly* (November 3, 1989): 52.

57. Wendy Kaminer, "Chances Are You're Co-Dependent Too," *New York Times Book Review* (February 11, 1990): 26

58. Gayle Feldman, "On the Road to Recovery with Prentice-Hall, Ballantine, et al," *Publishers Weekly* (November 3, 1989):52; S. Katz and A. Liu, *The Codependency Conspiracy* (New York: Warner Books, 1991), p. 16.

59. Bette S. Tallen, "Twelve Step Programs: A Lesbian Feminist Critique," *NWSA Journal* 2 (1990): 396.

60. Bette S. Tallen, "Twelve Step Programs: A Lesbian Feminist Critique," *NWSA Journal* 2 (1990): 404–405.

61. Bette S. Tallen, "Twelve Step Programs: A Lesbian Feminist Critique," *NWSA* Journal 2 (1990): 405.

Chapter 4

1. A. G. Britton, "Thin Is Out, Fit Is In," *American Health* (July/August 1988): 66–71; T. F. Cash, B. A. Winstead, and L. H. Janda, "The Great American Shape-up," *Psychology Today* (April 1986): 30–37.

2. R. Corliss, "Sexes: The New Ideal of Beauty," *Time* (August, 30, 1982): 72–73.

3. April Fallon, "Culture in the Mirror: Sociocultural Determinants of Body Image" in *Body Images: Development, Deviance and Change*, eds. Thomas F. Cash and Thomas Pruzinsky (New York: Guilford Press, 1990), p. 91.

4. Anthony Synnott, *The Body Social: Symbolism, Self and Society* (New York: Routledge,1993), p. 16.

5. Richard A. Gordon, *Anorexia and Bulimia: Anatomy of a Social Epidemic* (Cambridge, MA: Basil Blackwell, 1990), p. 96. Some researchers note how as a society Americans are obsessed with health—wellness as well as fitness—and that they equate these things with "personal salvation." Getting fit and staying healthy, these authors suggest, is becoming a moral imperative within our society and this imperative is used by groups and agencies to justify attempts to control those who do not measure up. They suggest there is a growing "health fascism" that exercises "increasing viligence and control over what people put into their bodies and what they put their bodies into" (p. 259). See: Charles Edgley and Dennis Brissett, "Health Nazis and the Cult of the Perfect Body: Some Polemical Observations,"*Symbolic Interaction* 13(1990):257–259.

6. Some research suggests that "compulsive exercise" is becoming a clinical issue: See: "S. Wichman and D. R. Martin," Exercise Excess: Treating Patients Addicted to Fitness," *The Physician and Sports Medicine* 20 (1992):193–200.

7. "The Fitness Industry—Snow Motion," *The Economist* 326(March 27, 1993): 71.

8. *The Lifestyle Market Analyst 1993: A Reference Guide for Consumer Market Analysis* (Wilmette, IL: Standard Rate and Data Service, 1993).

9. Census of Service Industries, 1982 (Washington, DC: U.S. Department of Commerce, 1982).

10. U.S. Industrial Outlook, 1993 (Washington, DC: U.S. Department of Commerce, January, 1993).

11. Cindee Miller, "Convenience, Variety, Spark Huge Demand for Home Fitness Equipment," *Marketing News* 6 (March 16, 1992): 2.

12. "The Fitness Industry—Snow Motion," *The Economist* 326 (March 27, 1993):71.

13. Monte Williams, "People to Watch," *Advertising Age* 61 (December 3, 1990):36.

14. Adrienne Ward, "Americans Step into a New Fitness Market," *Advertising Age* 61 (December 3, 1990):39.

15. "Exercise Video: Toned Up and Taking Off—Again," *Video Marketing News* (Potomac, MD: Phillips Business Information, 1992), pp. 61–63.

16. Personal conversation with industry analyst requesting anonymity (January 1994).

17. Catherine Applefield, "Keeping Up with All the Fondas," *Billboard* (November 16, 1991): 56.

18. "Exercise Video: Toned Up and Taking Off—Again," *VideoMarketing News* (Potomac, MD: Phillips Business Information, 1992), p. 58.

19. "The Fitness Industry—Snow Motion," *The Economist* 326 (March 27, 1993):72.

20. Chip Walker, "Fat and Happy," *American Demographics* 15(January 1993): 52–57.

21. Chip Walker, "Fat and Happy," *American Demographics* 15 (January 1993):52.

22. See: Stanley Wohl, *Medical Industrial Complex* (New York: Harmony Books, 1984), p. 18. Wohl notes: "Originally the term MIC (Medical Industrial Complex) was coined to describe the loose influential alliance of drug companies and medical associations. It was first appropriated for use in the present sense in 1980 by Dr. Arnold S. Relman, editor of the *New England Journal of Medicine*.

23. See: Douglas Shenson, "Will M.D." Mean "More Dollars?," *The New York Times*, Thursday, May 23, 1985): 27.

24. Donald W. Light, "Corporate Medicine for Profit," *Scientific American*, (December 1986):38.

25. Stanley Wohl, *The Medical Industrial Complex* (New York: Hamony Books, 1984), p. 3.

26. See Diana Dull, "Before and Afters: Television's Treatment of the Boom in Cosmetic Surgery." Paper Presented the 84th annual meeting of the American Sociological Association, San Francisco, CA (August 9—13, 1989).

27. Diana Dull, "Before and Afters: Television's Treatment of the Boom in Cosmetic Surgery." Paper Presented at the 84th annual meeting of the American Sociological Association, San Francisco, CA (August 9—13, 1989), p. 2.

28. American Academy of Facial Plastic and Reconstructive Surgery, *The Face Book: The Pro's and Con's of Facial Plastic and Reconstructive Surgery* (Washington, DC: Acropolis Books Ltd., 1988).

29. "The Price of Beauty," *The Economist* (January 11, 1992):25–26.

30. See: Sue Woodman, "Losing Fat Permanently," *Fitness* (March/ April 1994):38–39.

31. Nancy Pappas, "Body by Liposuction," *Hippocrates: The Magazine of Health and Medicine* 3 (May/June 1989):26–30. While this method of liposuction is still the most common, a new procedure has been developed called "tumescent techniques" or fluid infusion. The areas of skin are "injected with a diluted local anesthetic solution which bloats the tissue to almost twice its original thickness, allowing the surgeon to work closer to the skin surface thus minimizing the chance of uneven results." See: Sue Woodman, "Losing Fat Permanently," *Fitness* (March/ Arpil 1994), p. 34.

32. Gina Kolata, "Accord on Implant Suit Brings Flood of Inquiries," *New York Times* (September 11, 1993): 7.

33. Adriane Fugh-Berman, "Training Doctors to Care for Women," *Technology Review* (February/March), 1994):35.

34. See: Lisa Billowitz, "Breast Implants: In the Aftermath of Corporate Greed," *Sojourner* 17 (August 1992):12; Gina Kolata, "Details of Implant Settlement Announced by Federal Judge," *New York Times* (April 5, 1994):A1; Gina Kolata, "3

Companies Near Landmark Accord on Breast Implant Lawsuits," *New York Times* (March 24, 1994):B10.

35. Sue Woodman, "Losing Fat Permanently," *Fitness* (March/April, 1994):39.

36. Armour Forse and George L. Blackburn, "Morbid Obesity: Weighing Treatment Options." Unpublished paper. Nutrition/Metabolism Laboratory, Department of Surgery, New England Deaconess Hospital, Harvard Medical School, Boston, MA, 1989.

37. Lisa Schoenfielder and Barb Weiser, *Shadow on a Tightrope: Writings by Women on Fat Oppression* (Iowa City, IA: Aunt Lute Book Co., 1983), pp. 161, 10, 186.

38. Esther Rothblum notes that the potential side effects of stomach stapling were stated as follows: "leakage of stomach fluid into the abdomen, ventral hernia, potassium deficiency, urinary tract infection, anemia, vitamin deficiency, osteoporosis due to lack of calcium, diarrhea, constipation, vomiting, malnutrition, stomach cancer and death." Esther D. Rothblum, "Women and Weight: Fad and Fiction," *The Journal of Psychology* 124 (1990): 19.

Chapter 5

1. Anthropologist Mary Douglas was one of the first social scientists to point out the implications of the social meanings of the body: "The social body constrains the way the physical body is perceived. The physical experience of the body, always modified by the social categories through which it is known, sustains a particular view of society." Mary Douglas, *Natural Symbols: Explorations in Cosmology*, 2nd ed. (London: Barrie and Jenkins [1970], 1973), p. 93.

2. R. M. Lerner, S. A. Karabenick, and J. L. Stuart, "Relations Among Physical Attractiveness, Body Attitudes and Self-Concept in Male and Female College Students," *Journal of Psychology* (1973): 85, 19–129. P. Rozin & A. E. Fallon, "Body Image, Attitudes to Weight and Misperceptions of Figure Preferences of the Opposite Sex: A Comparison of Men and Women in Two Generations," *Journal of Abnormal Psychology* 97 (1988): 342–345. Linda A. Jackson, *Physical Appearance and Gender: Sociobiological and Sociocultural Perspectives* (Albany, NY: State University of New York Press. 1992).

3. George Herbert Mead, *Mind, Self and Society* (Chicago: University of Chicago Press, 1934), p. 135.

4. George Herbert Mead, *Mind, Self and Society* (Chicago: University of Chicago Press, 1934), p. 202.

5. Charles Horton Cooley, *Social Organization*, (New York: Schocken Books, 1962 [1909]).

6. G. R. Adams, "Physical Attractiveness Research," *Human Development* 20 (1977): 217–239 ; Marcia Millman, *Such a Pretty Face* (New York: Berkeley Books, 1980); J. Rodin, L. Silberstein, and R. Striegel-Moore, "Women and Weight: A Normative Discontent," in *Psychology and Gender: Nebraska Symposium on Motivation.* ed. T. B. Sonderegger (Lincoln: University of Nebraska Press, 1985), pp. 267—307; R. G. Simmons and F. Rosenberg, "Sex, Sex Roles, and Self-Image," *Journal of Youth and Adolescence* 4 (1975): 229–258.

7. This is not to imply that women passively adapt to the dictates of patriarchal/capitalist mirrors of beauty. They also act according to their consciousness and will.

We will argue, however, that there are strong rewards and punishments that women experience in the process of growing up that serve to provide women with a "pseudo-choice." They choose to conform and their actions combine over space and time to recreate or reproduce the beauty standards of patriarchy and capitalism.

8. Thomas F. Cash, "The Psychology of Physical Appearance: Aesthetics, Attributes and Images," in *Body Images Development, Deviance and Change*, eds. Thomas F. Cash and Thomas Pruzinsky (New York: Guilford Press, 1990), p. 53. See also: K. K. Dion, E. Berscheid, and E. Walster, "What Is Beautiful Is Good," *Journal of Personality and Social Psychology* 24 (1972):285–290. For a review of the literature on this topic see: E. Hatfield and S. Sprecher, *Mirror, Mirror: The Importance of Looks in Everyday Life* (Albany: State University of New York Press, 1986).

9. For example, research on infants notes that the perception of cuteness is related to the degree to which maternal bonding takes place. See: G. H. Elder, Jr., T. V. Nguyen, and A. Caspi, "Linking Family Hardship to Children's Lives," *Child Development* 56 (1985):361–375.

10. Thomas F. Cash, "The Psychology of Physical Appearance: Aesthetics, Attributes, and Images," in *Body Images: Development, Deviance, and Change*, eds. Thomas F. Cash and Thomas Pruzinsky (New York: Guilford Press, 1990), p. 54; L. Berkowitz and A. Frodi, "Reactions to a Child's Mistakes as Affected by His/her Looks and Speech," *Social Psychology Quarterly* 42(1979): 420–425; K. K. Dion, "Physical Attractiveness and Evaluation of Children's Transgressions," *Journal of Personality and Social Psychology* 24 (1972):207–213; K. K. Dion, "Children's Physical Attractiveness and Sex as Determinants of Adult Punitiveness," *Developmental Psychology* 10 (1974): 772–778; V. McCabe, "Facial Proportions, Perceived Age, and Caregiving," in *Social and Applied Aspects of Perceiving Faces*, ed. T. R. Alley (Hillsdale, NJ: Erlbaum, 1988), pp. 89—95.

11. K. K. Dion, "Physical Attractiveness and Evaluation of Children's Transgressions," *Journal of Personality and Social Psychology* 24 (1972): 211–212.

12. E. Berscheid and E. Walster, "Beauty and the Beast," *Psychology Today* (October, 1972):45.

13. W. H. Sheldon, *The Varieties of Human Physique: An Introduction to Constitutional Psychology* (New York: Harper & Row, 1940). While the relationship between body type and personality characteristics was not empirically supported in subsequent research studies for given individuals, there is some evidence that others believe that this relationship is true, even when there is no empirical proof. For a review of this research see: Linda A. Jackson, *Physical Appearance and Gender: Sociobiological and Sociocultural Perspectives* (Albany: State University of New York Press. 1992), pp. 156–158.

14. Rita Freedman, *Beauty Bound* (Lexington, MA: D.C. Heath & Co., 1986). One newspaper reporter commented on the coming of the "New Puritans." and notes: "New Puritans are not merely concerned with developing clearer complexions or trimmer thighs. They pursue self-denial as an end in itself, out of an almost mystical belief in the purity it confers. They work hard and play harder—if your idea of play is Olympic competition. . . . But while the benefits of moderate exercise and the value of a high-fiber low-cholesterol diet are enshrined in the canons of medicine and the annual reports of the publishing world, and while it is probably a good idea to use a little caution in choosing a sex partner, the New Puritanism goes well beyond the minimum daily requirements for sound minds in sound bodies." (p. 26). See: Dinitia Smith, "The New Puritans: Deprivation Chic," *New York Magazine* (June 11, 1984): 24–29.

15. G. R. Adams, "Physical Attractiveness Research," *Human Development* 20

(1977):217–240; R. M. Lerner and S. A. Karabenick, "Physical Attractiveness, Body Attitudes, and Self-Concept in Late Adolescents," *Journal of Youth and Adolescence* 3 (1974): 307–316; R. G. Simmons and F. Rosenberg, "Sex, Sex Roles, and Self-Image," *Journal of Youth and Adolescence* 4 (1975): 229–258.

16. S. A. Richardson, N. Goodman, A. H. Hastorf, and S. M. Dornbusch, "Cultural Uniformity in Reaction to Physical Disabilities," *American Sociological Review* 26 (1961):241–247.

17. B. E. Vaughn and J. H. Langolis, "Physical Attractiveness as a Correlate of Peer Status and Social Competence in Pre-school Children," *Developmental Psychology* 19 (1983): 561–567.

18. See: E. Berscheid and E. Walster, "Beauty and the Beast," *Psychology Today* (October, 1972): 42–46. For an excellent review of this literature, see: J. Rodin, L. Silberstein, and R. Streigel-Moore, "Women and Weight: A Normative Discontent," in *Psychology and Gender: Nebraska Symposium on Motivation*, ed. T. B. Sonderegger (Lincoln: University of Nebraska Press, 1985), pp 267–307.

19. M. Tiggemann and E. D. Rothblum, "Gender Differences in Social Consequences of Perceived Overweight in the United States and Australia," *Sex Roles* 18 (1988): 75–86; J. Stake and M. L. Lauer, "The Consequences of Being Overweight: A Controlled Study of Gender Differences," *Sex Roles* 17 (1986): 31–47.

20. D. Bar-Tal and L. Saxe, "Physical Attractiveness and Its Relationship to Sex-Role Stereotyping," *Sex Roles* 2 (1976): 123–133; P. W. Blumstein and P. Schwartz, *American Couples* (New York: Morrow, 1983).

21. Questionnaires were distributed to 960 sophomores at a private New England college. 395 questionnaires were returned and analyzed. The resulting response rate was 41% and the sample consisted of 71% females and 29% males.

22. Our research is confirmed by numerous studies of gender differences in perceived weight. See: S. Gray, "Social Aspects of Body Image: Perceptions of Normality of Weight and Affect on College Undergraduates," *Perceptual and Motor Skills* 10 (1977): 503–516.; A. Fallon and P. Rozin, "Sex Differences in Perceptions of Desirable Body Shape," *Journal of Abnormal Psychology* 94 (1985): 102–105; P. Rozin and A. Fallon, "Body Image, Attitudes to Weight and Misperceptions of Figure Preference of the Opposite Sex: A Comparison of Men and Women in Two Generations," *Journal of Abnormal Psychology* 97 (1988): 342–345; Linda A. Jackson, *Physical Appearance and Gender: Sociobiological and Sociocultural Perspectives* (Albany: State University of New York Press, 1992).

23. Paul S. Entmacher, M.D., Metropolitan Life Insurance Company's vice president and chief medical director was quoted in *Vogue* magazine concerning his perspective on his company's ideal weight chart. These tables "do not necessarily indicate the weights that reduce the likelihood of illness. Nor are the weights those at which people may look best." He is stressing that the weight charts are desirable in that they are associated with lowest mortality. See *Vogue* (September, 1983):706.

24. An elaborate analytical procedure involving students' self-reported weight and height and stated desired weights was used to determine the weight model ("cultural" or "medical") students were following. A medium size body frame was assumed in our calculations.

25. L. K. G. Hsu, "Classification and Diagnosis of the Eating Disorders," in *The Eating Disorders: Medical and Psychological Basis of Diagnosis and Treatment*, eds. B. J. Blinder, B. F. Chaitin, and R. S. Goldstein (New York : PMA Publishing, 1988), pp. 235—238.

Chapter 7

1. Excerpt from *Keep Young and Beautiful* by Harry Warren and Al Dubin and performed by Annie Lennox. Copyright 1933 (renewed) Warner Bros., Inc. All rights reserved. Used by permission of Warner Bros. Publications Inc., Miami, FL 33014

2. See: A.E. Anderson, "Anorexia Nervosa and Bulimia in Adolescent Males," *Pediatric Annals* 12 (1984) 901–904,907.

3. K. M. Bemis, "Current Approaches to the Etiology and Treatment of Anorexia Nervosa," *Psychology Bulletin* 85 (1978) 593–617 ; H. Bruch, *Eating Disorders: Obesity, Anorexia and the Person Within* (New York: Basic Books, 1973).

4. H. Bruch, *Eating Disorders: Obesity, Anorexia and the Person Within* (New York: Basic Books, 1973); A.H. Crisp, "Some Aspects of the Evolution Presentation and Follow-up of Anorexia Nervosa," *Proceedings of the Royal Society of Medicine* 58(1965): 814–820; H. G. Morgan and G. F. M. Russel, "Value of Family Background and Clinical Features as Predictors of Long-Term Outcome in Anorexia Nervosa: Four-year Follow-up Study of 41 Patients," *Psychological Medicine* 5 (1975): 355–371.

5. Even this number may be conservative; the real incidence is underreported because of the stigma attached to having an eating disorder.

6. The Eating Attitudes Test (EAT) was used as a measure for eating disorders. The EAT is a clinical test that has been used in other research with student populations. The test consists of 26 items designed to evaluate a broad range of behaviors and contains three scales: diet, bulimia, and oral control. The diet scale relates to an avoidance of fattening foods and a preoccupation with being thin. The bulimia scale consists of items reflecting a preoccupation with food as well as thoughts indicating bulimic behaviors. The third measure—oral control—relates to self-control about food. See: D. M Garner, M. P. Olmsted, Y. Bohr, and P. E. Garfinkel, "The Eating Attitudes Test: Psychometric Features and Clinical Correlates," *Psychological Medicine* 12(1982):871–878.

7. See: R. A. Gordon, "A Sociocultural Interpretation of the Current Epidemic of Eating Disorders," in *The Eating Disorders: Medical and Psychological Basis of Diagnosis and Treatment*, eds. B. J. Binder, B. F. Chaitin, and R. S. Goldstein (New York: PMA Publishing Company, 1988), pp. 151–163.

8. Sharlene Hesse-Biber, "Eating Patterns and Disorders in a College Population: Are College Women's Eating Problems a New Phenomenon?," *Sex Roles* 20: (1989):71–89; Sharlene Hesse-Biber, "Women, Weight and Eating Disorders: A Socio-Cultural and Political-Economic Analysis," *Women's Studies International Forum* 14 (1991):173–191.

9. D. M. Garner, M. S. Olmsted, and P. E. Garfinkel, "Does Anorexia Nervosa Occur on a Continuum? Subgroup of Weight-Preoccupied Women and Their Relationship to Anorexia Nervosa," *International Journal of Eating Disorders* 2 (1983): 11–20; E. J. Button and A. Whitehouse, "Subclinical Anorexia Nervosa," *Psychological Medicine* 11 (1981):509–516; Sharlene Hesse-Biber, "Report on a Panel Longitudinal Study of College Women's Eating Patterns and Eating Disorders: Noncontinuum Versus Continuum Measures," *Health Care for Women International* 13 (1992): 375–391.

10. J. Polivy and C. P. Herman, "Dieting and Binging: A Causal Analysis," *American Psychologist*, 40 (1985):193–201.

11. There is disagreement about the order of causality. It is difficult to ascertain

whether depression leads to eating issues or problematic eating issues leads to depression. See: W. J. Swift, D. Andrews, and N. E. Barklage, "The Relation Between Affective Disorder and Eating Disorders: A Review of the Literature," *American Journal of Psychiatry* 143(1986):290–299. Some researchers suggest that the onset of depression and eating issues in young women is a result poor body image. See: M. McCarthy, "The Thin Ideal, Depression, and Eating Disorders in Women," *Behavioral Research and Therapy* 28(1990):205–215.

12. See: L. R. Furst and P. W. Graham (eds.), *Disorderly Eaters: Texts in Self-Empowerment* (University Park: The Pennsylvania State University Press, 1992). See also: Becky Thompson, "Food, Bodies, and Growing Up Female: Childhood Lessons about Culture, Race, and Class, in *Feminist Perspectives on Eating Disorders*, eds. P. Fallon, M. A. Katzman, and S. C. Wooley (New York: Guilford Press, 1994), pp. 355–378. See also: M. L. Lawrence (ed), *Fed Up and Hungry: Women, Oppression and Food* (New York: Peter Bedrick Books, 1987).

13. Nancy Chodorow, *Feminism and Psychoanalytic Theory* (New Haven: Yale University Press, 1989), p. 43.

14. Stephen Wonderlich, "Relationship of Family and Personality Factors in Bulimia," in *The Etiology of Bulimia Nervosa: The Individual and Familial Context*, eds. J. H. Crowther, D. L. Tennenbaum, S. E. Hobfoll, and M. A. P. Stephens (London: Hemisphere Publishing Corporation, 1992), pp. 103–126. This review of the literature on family and personality factors in bulimia cites three specific types of families that are at high risk for daughters developing bulimia: (1) The perfect, (2) the overprotective, and (3) the chaotic. All three family types have "boundary" problems and stress the importance of weight and body image. All three types are characterized by "extreme levels of paternal (versus maternal) power" (p. 105). All three family types "reflect the difficulty that the family experiences negotiating the affected children's transition from adolescence to young adulthood." Barbara's family appears to be an example of the "perfect" family.

15. I. Attie and J. Brooks-Gunn, "Developmental Issues in the Study of Eating Problems and Disorders," in *The Etiology of Bulimia Nervosa: The Individual and Familial Context*, eds. J. H. Crowther, D. L. Tennenbaum, S. E. Hobfoll, and M. A. P. Stephens (London: Hemisphere Publishing Corporation, 1992), pp. 43–45. See also: Suzanne Alexander, "Egged on by Moms, Many Teen-agers Get Plastic Surgery," *The Wall Street Journal* (September 24, 1990): 1. For a good discussion of mother-daughter competition see especially: I. Attie and J. Brooks-Gunn, "Developmental Issues in the Study of Eating Problems and Disorders," in *The Etiology of Bulimia Nervosa: The Individual and Familial Context*, eds. J. H. Crowther, D. L. Tennenbaum, S. E. Hobfoll, and M. A. P. Stephens (London: Hemisphere Publishing Corporation, 1992), pp. 35–58; J. Rodin, R. H. Striegel-Moore, and L. R. Silberstein, "Vulnerability and Resilience in the Age of Eating Disorders," in *Risk and Protective Factors in the Development of Psychopathology*, eds. J. Rolf, A. Masten, D. Cicchetti, et al. (Cambridge: Cambridge University Press, 1990), pp. 366–390.

16. E. J. Button and A. Whitehouse, "A Subclinical Anorexia Nervosa," *Psychological Medicine* 11(1981) 509–516; D. M. Garner and P. E. Garfinkel, "The Eating Attitudes Test: An Index of the Symptoms of Anorexia Nervosa," *Psychological Medicine* 9 (1979): 273–279; M. Tamburrino, K. N. Franco, G. A. A. Bernal, B. Carroll, and A. J. McSweeny, "Eating Attitudes in College Students," *Journal of the American Medical Women's Association* 42 (1987):45–50; M. G. Thompson and D. Schwartz,

"Life Adjustment of Women with Anorexia Nervosa and Anorexic-like Behavior," *International Journal of Eating Disorders* 1(1982): 47–60.

17. J. J. Gray and K. Ford, "The Incidence of Bulimia in a College Sample," *International Journal of Eating Disorders* 4 (1985):201–210; K. A. Halmi, J. R. Falk, and E. Schwartz, "Binge-eating and Vomiting: A Survey of a College Population," *Psychological Medicine* 11 (1981): 697–706; K. J. Hart and T. H. Ollendick, "Prevalence of Bulimia in Working and University Women," *American Journal of Psychiatry* 142 (1985) 851–854; M. A. Katzman, S. A. Wolchik, and S. L. Braver, "The Prevalence of Frequent Binge Eating and Bulimia in a Nonclinical College Sample," *International Journal of Eating Disorders* 3 (1984): 53–62; R. L. Pyle, P. A. Halvorson, P. A. Neuman, and J. E. Mitchell, "The Increasing Prevalence of Bulimia in Freshman College Students," *International Journal of Eating Disorders* 5 (1986): 631–647; R. L. Pyle, J. E. Mitchell, E. D. Eckert, P. A. Halvorson, P. A. Neuman and G. M. Goff, "The Incidence of Bulimia in Freshman College Students," *International Journal of Eating Disorders* 2 (1983): 75–85.

18. R. C. Hawkins and P. F. Clement, "Development and Construct Validation of a Self-Report Measure of Binge-eating Tendencies," *Addictive Behaviors* 5 (1980): 219–226; K. A. Halmi, J. R. Falk, and E. Schwartz, "Binge-Eating and Vomiting: A Survey of a College Population," *Psychological Medicine* 11 (1981) 697–706; P. A. Ondercin, "Compulsive Eating in College Women," *Journal of College Student Personnel* 19 (1979): 153–157; D. M. Zuckerman, A. C. Colby, N. C. Ware, and J. S. Lazerson, "The Prevalence of Bulimia Among College Students," *American Journal of Public Health.* 76(1986): 1135–1137.

19. See: M. Boskind-Lodahl and W. C. White, Jr., "The Definition and Treatment of Bulimarexia in College Women," *Journal of the American College Health Association* 27 (1979): 84–86, 97.

20. R.Striegel-Moore, L. Silberstein, and J. Rodin, "Toward Understanding the Risk Factors for Bulimia," *American Psychologist* 41(1986): 246–263; S. Squire, *The Slender Balance: Causes and Cures for Bulimia, Anorexia, and the Weight-Loss/Weight-Gain Seesaw* (New York: Putnam, 1983).

21. S. M. Dornbusch, J. M. Carlsmith, P. D. Duncan, R. T. Gross, J. A. Martin, P. L. Ritter, and B. Siegel-Gorelick, "Sexual Maturation, Social Class, and the Desire to be Thin Among Adolescent Females," *Developmental and Behavioral Pediatrics* 5 (1984): 308–314; A. J. Stunkard, E. E. d'Aquili, S. Fox, and R. D. L. Filion, "Influence of Social Class on Obesity and Thinness in Children," *Journal of the American Medical Association* 221 (1972): 579–584.

22. See: R. Striegel-Moore, L. Silberstein, and J. Rodin,"Toward Understanding the Risk Factors for Bulimia," *American Psychologist* 41(1986): 246–263.

23. C. S. Crandall, "Social Contagion of Binge Eating," *Journal of Personality and Social Psychology* 55 (1988): 588–598.

24. C. S. Crandall, "Social Contagion of Binge Eating," *Journal of Personality and Social Psychology* 55 (1988): 588–598.

25. J. Rodin, L. Silberstein, and R. Striegel-Moore, "Women and Weight: A Normative Discontent," in *Psychology and Gender: Nebraska Symposium on Motivation*, ed. T. B. Sonderegger (Lincoln: University of Nebraska Press, 1985), pp. 267–307.

26. Dorothy C. Holland and Margaret A. Eisenhart, *Educated in Romance: Women, Achievement, and College Culture*, (Chicago:University of Chicago Press, 1990), p. 201.

27. See: M. F. Hovell, C. R. Mewborn, Y. Randle, and J. S. Fowler-Johnson,

"Risk of Excess Weight Gain in University Women: A Three-year Community Controlled Analysis," *Addictive Behaviors* 10 (1985): 15–28; R. H. Striegel-Moore, L. R. Silberstein, P. Frensch, and J. Rodin, "A Prospective Study of Disordered Eating Among College Students," *International Journal of Eating Disorders* 8 (1989): 499–509.

28. S. Squire, *The Slender Balance: Causes and Cures for Bulimia, Anorexia, and the Weight-loss/Weight-gain Seesaw* (New York: Putnam, 1983). One student at Boston College commented to me that at the Massachusetts Institute of Technology, the weight gain of freshman women was known as the "MIT 15." She commented on the tremendous pressure on women to succeed in a traditionally male science and engineering school.

29. There is debate about how much weight gain women experience and some researchers note that for some it may be more of a myth of college life than a reality. See: C. N. Hodge, L.A. Jackson, and L. A. Sullivan, "The 'Freshman 15': Facts and Fantasies About Weight Gain in College Women," *Psychology of Women Quarterly* 17 (1993):119–126.

30. M. Brouwers, "Depressive Thought Content Among Female College Students with Bulimia," *Journal of Counseling and Development* 66 (1988):425–428; P. R. Holleran, J. Pascale, and J. Fraley. "Personality Correlates of College Age Bulimics," *Journal of Counseling and Development* 66 (1988): 378–381; L. P. F. McCanne, "Correlates of Bulimia in College Students: Anxiety, Assertiveness, and Locus of Control," *Journal of College Student Personnel* (1985):306–310; L. B. Mintz and N. E. Betz, "Prevalence and Correlates of Eating Disordered Behaviors Among Undergraduate Women," *Journal of Counseling Psychology* 35 (1988): 463–471; M. Pertschuk, M. Collins, J. Kreisberg, and S. S. Fager, "Psychiatric Symptoms Associated with Eating Disorders in a College Population," *International Journal of Eating Disorders* 5 (1986): 563–568; J. Rodin, L. Silberstein, and R. Striegel-Moore, "Women and Weight: A Normative Discontent," *Psychology and Gender: Nebraska Symposium on Motivation*, ed. T. B. Snoderegger (Lincoln: University of Nebraska Press, 1985), pp. 267–307; S. A. Segal and C. B. Figley, "Bulimia: Estimate of Increase and Relationship to Shyness," *Journal of College Student Personnel* 26 (1985):240–244; B. Silverstein and L. Perdue, "The Relationship Between Role Concerns, Preferences for Slimness, and Symptoms of Eating Problems Among College Women," *Sex Roles* 18 (1988):101–106; R. Striegel-Moore, L. Silberstein, and J. Rodin, "Toward Understanding of the Risk Factors for Bulimia," *American Psychologist* 41 (1986):246–263.

31. S. Hesse-Biber and M. Marino, "From High School to College: Changes in Women's Self-Concept and Its Relationship to Eating Problems," *The Journal of Psychology* 125 (1991): 199–216.

32. J. Freeman, "How to Discriminate Against Women Without Really Trying," in *Women: A Feminist Perspective*, ed. Jo Freeman (Palo Alto, CA: Mayfield, 1975).

33. J. Freeman, "How to Discriminate Against Women Without Really Trying," in *Women: A Feminist Perspective*, ed. Jo Freeman (Palo Alto, CA: Mayfield, 1975), p. 216.

Chapter 8

1. Statistics and quote are from Aimee Pohl, "Teen Magazine's Message to Girls: You Can Be Anything . . . Except Yourself," *Extra!: A Publication of FAIR* (Fairness and Accuracy in Reporting) (New York: FAIR/EXTRA!, 1992): 28.

2. L. M. Mellin, S. Scully, and C. E. Irwin, "Disordered Eating Characteristics in Preadolescent Girls." Paper presented at American Dietetic Association annual meeting, Las Vegas, Nevada (October 28, 1986); D. M. Stein and P. Reichert, "Extreme Dieting Behaviors in Early Adolescence," *Journal of Early Adolescence* 10 (1990): 108–121.

3. R. H. Striegel-Moore, "Prevention of Bulimia Nervosa: Questions and Challenges," in *The Etiology of Bulimia Nervosa: The Individual and Familial Context*, eds. Janis H. Crowther, Daniel L. Tennenbaum, Stevan E. Hobfoll, and Mary Ann Parris Stephens (Washington, DC: Hemisphere Publishing Corporation, 1992), pp. 203–223.

4. R. H. Striegel-Moore, "Prevention of Bulimia Nervosa: Questions and Challenges," in *The Etiology of Bulimia Nervosa: The Individual and Familial Context*, eds. Janis H. Crowther, Daniel L. Tennenbaum, Stevan E. Hobfol,l and Mary Ann Parris Stephens (Washington, DC: Hemisphere Publishing Corporation, 1992), p. 212.

5. I. Attie and J. Brooks-Gunn, "Weight Concerns as Chronic Stressors in Women," in *Gender and Stress*, eds. R. C. Barnett, L. Biener, and G.K. Baruch (New York: The Free Press, 1987), p. 233. See also: B. A. Hamburg, "Early Adolescence as a Life Stress," in *Coping and Health*, eds. Seymour Levine and Holger Ursin (New York: Plenum, 1980). Pre-teens and teens are appearing more frequently in advertising. The rationale for this is that teens are more influenced by advertising in which other teens are selling products. R. H. Striegel-Moore, "Prevention of Bulimia Nervosa: Questions and Challenges," in *The Etiology of Bulimia Nervosa: The Individual and Familial Context*, eds. Janis H. Crowther, Daniel L. Tennenbaum, Stevan E. Hobfoll, and M. A. P. Stephens (Washington, DC: Hemisphere Publishing Corporation, 1992), p. 213.

6. Jenifer Davis and Robert Oswalt, "Societal Influences on a Thinner Body Size in Children," *Perceptual and Motor Skills* 74 (1992): 697–698.

7. Jules Hirsch and Jerome L. Knittle, "Effect of Early Nutrition on the Development of Rat Epididymal Fat Pads," *Journal of Clinical Investigation* 47 (1968): 2091–2098 and J. Hirsch, J. Knittle, "Cellularity of Obese and Nonobese Human Adipose Tissue," *Federation Proceedings* 29 (1970), 1516–1521. As cited (p. 172) in Roberta Pollack Seid, *Never Too Thin: Why Women Are at War with Their Bodies* (New York: Prentice-Hall Press, 1989).

8. Roberta Pollack Seid, *Never Too Thin: Why Women Are at War with Their Bodies*, (New York: Prentice-Hall Press, 1989), p. 172.

9. Roberta Pollack Seid, *Never Too Thin: Why Women Are at War with Their Bodies* (New York: Prentice-Hall Press, 1989), p. 172.

10. See: Jean Mayer, "Fat Babies Grow into Fat People," *Family Health* 5 (1973): 24–38; Jean Mayer, "When to Start Dieting? At Birth," *Medical World News* (September 1973): 31–33. Cited in Roberta Pollack Seid, *Never Too Thin: Why Women Are at War with Their Bodies*. (New York: Prentice-Hall Press, 1989), p. 173.

11. Roberta Pollack Seid, *Never Too Thin: Why Women Are at War with Their Bodies* (New York: Prentice-Hall Press, 1989), p. 173.

12. Roberta Pollack Seid, *Never Too Thin: Why Women Are at War with Their Bodies*, (New York: Prentice-Hall Press, 1989), p. 173.

13. W. Feldman, E. Feldman, and J. T. Goodman, "Culture Versus Biology: Children's Attitudes Toward Thinness and Fatness," *Pediatrics* 81(1988):190.

14. I. Attie and J. Brooks-Gunn, "Weight Concerns as Chronic Stressors in Women," in *Gender and Stress*, eds. R. C. Barnett, L. Biener, and G. K. Baruch (New

York: Free Press, 1987), p. 228; R. M. Lerner, "Children and Adolescents as Producers of Their Development," *Developmental Review* 2 (1982):342–370.

15. L. M. Mellin, S. Scully, and C. E. Irwin, "Disordered Eating Characteristics in Preadolescent Girls." Paper presented at American Dietetic Association annual meeting, Las Vegas, Nevada (October 28, 1986), Table 2.

16. L. M. Mellin, S. Scully, and C. E. Irwin, "Disordered Eating Characteristics in Preadolescent Girls." Paper presented at American Dietetic Association annual meeting, Las Vegas, Nevada (October 28, 1986), p. 6.

17. L. M. Mellin, S. Scully, and C. E. Irwin, "Disordered Eating Characteristics in Preadolescent Girls." Paper presented at American Dietetic Association annual meeting, Las Vegas, Nevada (October 28, 1986).

18. See W. Feldman, E. Feldman, and J.T. Goodman, "Culture Versus Biology: Children's Attitudes Toward Thinness and Fatness," *Pediatrics* 81 (1988): 190–194.

19. W. Feldman, E. Feldman, and J. T. Goodman, "Culture Versus Biology: Children's Attitudes Toward Thinness and Fatness," *Pediatrics* 81 (1988) :190.

20. Laurel Mellin, Sarah Scully, and Charles E. Irwin, "Disordered Eating Characteristics in Preadolescent Girls." Paper presented at the American Dietetic Association annual meeting, Las Vegas, Nevada (October 28, 1986).

21. For a discussion of the research literature on "fear of fat" in young children, see: M. H. Thelen, C. M. Lawrence, and A. L. Powell, "Body Image, Weight Control and Eating Disorders Among Children," in *The Etiology of Bulimia Nervosa: The Individual and Familial Context*, eds. J. H. Crowther, D. L. Tennenbaum, S. E. Hobfoll, and M. A. P. Stephens (London: Hemisphere Publishing Co., 1992.), pp.81–101. A note of caution must be voiced here. Some researchers note that while teenage girls may talk about fear of fat and dieting, these attitudes may not translate into severe dieting behaviors as some research on adolescents suggests. A recent study by Mark Nichter and Mimi Nichter questioned adolescents about their dieting behavior by asking "What does being on a diet mean?" They noted that for teens being on a diet "often constitutes a ritual activity wherein the consumption of token foods is suspended" (p. 264). Mark Nichter and Mimi Nichter, "Hype and Weight," *Medical Anthropology* 13, (1991):249–284. "Fat talk" among adolescents has important consequences even if these attitudes do not readily translate into severe dieting. Researchers Mimi Nichter and Nancy Vuckovic's longitudinal study of adolescent teens' "fat talk" suggests that: "By engaging in fat talk, females present themselves to others as responsible being concerned about their appearance. . . . Irrespective of what actions girls are taking to achieve their body goals, they are attempting to reproduce the cultural ideal through their discourse" (p. 127). See: Mimi Nichter and Nancy Vuckovic, "Fat Talk: Body Image Among Adolescent Girls," in *Many Mirrors: Body Image and Social Relations*, ed. by Nicole Sault (New Brunswick, NJ: Rutgers University Press, 1994), pp. 109–131.

22. N. Moses, M. Banilivy, and F. Lifshitz, "Fear of Obesity Among Adolescent Girls." *Pediatrics* 83 (1989): 393–398. Other research studies reveal similar results. For an excellent review of the literature on eating disorders among children see: M. H. Thelen, C. M. Lawrence, and A. L. Powell, "Body Image, Weight Control and Eating Disorders Among Children," in *The Etiology of Bulimia Nervosa: The Individual and Familial Context*, eds. J. H. Crowther, D. L. Tennenbaum, S. E. Hobfoll, and M. A. P. Stephens (London: Hemisphere Publishing Co., 1992), pp. 81–101.

23. F. Lifshitz, N. Moses, C. Cervantes et al., "Nutritional Dwarfing in Adolescents," *Seminar on Adolescent Medicine* 3 (1987):255–266, F. Lifshitz and N. Moses,

"Nutritional Dwarfing: Growth, Dieting, and Fear of Obesity," *Journal of the American College of Nutrition* 7 (1988): 367–376.

24. "The Littlest Dieters," *Newsweek* (July 27, 1987): 48.

25. F. Lifshitz , N. Moses, C. Cervantes, and L. Ginsberg, "Nutritional Dwarfing in Adolescents," *Seminars in Adolescent Medicine* 3 (1987): 255–256; D. D. Marino and J. C. King, "Nutritional Concerns During Adolescence," *Pediatric Clinics of North America* 27 (1980): 125–137.

26. I. Attie and J. Brooks-Gunn, "Weight Concerns as Chronic Stressors in Women," in *Gender and Stress*, eds. R. C. Barnett, L. Biener, and G. K. Baruch (New York: Free Press, 1987), pp. 218–254. See especially in M. P. Levine and L. Smolak, "Toward a Model of the Developmental Psychopathology of Eating Disorders: The Example of Early Adolescence," in *The Etiology of Bulimia Nervosa: The Individual and Familial Context*, eds. J. H. Crowther, D. L. Tennenbaum, S. E. Hobfoll, and M.A. P.. Stephens (London: Hemisphere Publishing Corp., 1992), pp. 69–80.

27. D. D. Marino and J. C. King, "Nutritional Concerns During Adolescence," *Pediatric Clinics of North America* 27 (1980): 125–137; M. P. Warren, "Physical and Biological Aspects of Puberty," in *Girls at Puberty: Biological and Psychosocial Perspectives*, eds. J. Brooks-Gunn and A. C. Petersen (New York: Plenum, 1983). A certain amount of fat on the body is required for reproduction. Adolescent girls undergo an increase in fat around puberty.

28. M. Millman, *Such a Pretty Face: Being Fat in America* (New York: Berkeley, 1980), p. 216.

29. M. Millman, *Such a Pretty Face: Being Fat in America* (New York: Berkeley, 1980), p. 218.

30. M. Millman, *Such a Pretty Face: Being Fat in America* (New York: Berkeley, 1980), p. 224.

31. C. J. Nemeroff, R. I. Stein, N. S. Diehl, K.M. Smilack, "From the Cleavers to the Clintons: Role Choices and Body Orientation as Reflected in Magazine Article Content," *International Journal of Eating Disorders* 16 (1994):167–176.

32. See: M. E. Mishkind, J. Rodin, L. R. Silberstein, and R. H. Striegel-Moore, "The Embodiment of Masculinity: Cultural, Psychological, and Behavioral Dimensions," *American Behavioral Scientist* 29 (1986):545–562. See also: "You're So Vain," *Newsweek* (April 14, 1986):48–55.

33. J. Ogden, *Fat Chance! The Myth of Dieting Explained* (New York: Routledge, 1992).

34. Data cited in A. Synnott, *The Body Social: Symbolism, Self and Society* (New York: Routledge, 1993), p. 32. This research comes from two separate studies. One study conducted in 1972 is from E. Berscheid, E. Walster, and G. Bohrnstedt, "The Happy American Body: A Survey Report," *Psychology Today* 7 (1973): 119–131, and a study conducted by T. F. Cash, B. A. Winstead, and L. H. Janda, "The Great American Shape-Up: Body Image Survey Report," *Psychology Today* (April 1986), pp. 30–37.

35. T. F. Cash, "The Psychology of Physical Appearance: Aesthetics, Attributes, and Images," in *Body Images: Development, Deviance and Change*, eds. T. G. Cash and T. Pruzinsky (New York: Guildord Press, 1990), pp. 51–79. Some controversy exists surrounding the extent to which men are dissatisfied with their bodies. In a study of college students' perceptions of their body weight, one researcher found that "men judged their relative weight more accurately and generally felt that their current body shape was very close to what women wanted in an ideal man. Thus, men's perceptions

help keep them satisfied with their bodies, whereas women's perceptions motivate them toward weight obsession and dieting" (p. 93). April Fallon, "Culture in the Mirror: Sociocultural Determinants of Body Image," in *Body Images: Development, Deviance and Change*, eds. by Thomas F. Cash and Thomas Pruzinsky. (New York: Guilford Press, 1990), pp. 80–109.

36. M E. Mishkind, J. Rodin, L. R. Silberstein, and R. H. Striegel-Moore, "The Embodiment of Masculinity: Cultural, Psychological , and Behavioral Dimensions," *American Behavioral Scientist* 29 (1986): 555. See also: "You're So Vain," *Newsweek* (April 14, 1986):48–55. C. Edgley and D. Brissett, "Health Nazis and the Cult of the Perfect Body: Some Polemical Observations," *Symbolic Interaction* 13 (1990):257–279.

37. D. MacCannell and J. F. MacCannell, "The Beauty System," in *The Ideology of Conduct: Essays in Literature and the History of Sexuality*, eds. N. Armstrong and L. Tennenhouse (New York: Methuen, 1987), p. 207.

38. J. Rodin, *Body Traps: Breaking the Binds That Keep You From Feeling Good About Your Body* (New York: William Morrow and Co., Inc., 1992), pp. 38–39.

39. A. Kimbrell, "Body Wars: Can the Human Body Survive the Age of Technology?," *Utne Reader* (May/June 1992): 53.

40. R. P. Seid, *Never Too Thin: Why Women Are at War with their Bodies* (New York: Prentice-Hall, 1989), p. 116.

41. A. Drewnowski and D. K. Yee, "Men and Body Image: Are Males Satisfied with Their Body Weight?," *Psychosomatic Medicine* 49 (1987): 626–634.

42. J. Rodin, *Body Traps: Breaking the Binds That Keep You From Feeling Good About Your Body*, (New York: William Morrow and Co., Inc., 1992), p. 181.

43. R.A. Gordon, *Anorexia and Bulimia: Anatomy of a Social Epidemic* (Cambridge, MA: Basil Blackwell, Inc, 1990), p. 32.

44. J. Rodin, *Body Traps: Breaking the Binds That Keep You From Feeling Good About Your Body* (New York: William Morrow and Co., Inc., 1992), p. 36.

45. J. Rodin, *Body Traps: Breaking the Binds That Keep You From Feeling Good About Your Body* (New York: William Morrow and Co., Inc., 1992), p. 88.

46. Harry Gwirtsman, cited in:"Bulimia: Not for Women Only," (no author), *Psychology Today* (March 1984): 10.

47. This story is taken from: Gabriella Stern, "The Anorexic Man Has Good Reason To Feel Neglected," *Wall Street Journal* (October 18, 1993): A1.

48. Gabriella Stern, "The Anorexic Man Has Good Reason To Feel Neglected," *Wall Street Journal* (October 18, 1993):A1.

49. Gabriella Stern, "The Anorexic Man Has Good Reason To Feel Neglected." *Wall Street Journal* (October 18, 1993):A1.

50. Gabriella Stern, "The Anorexic Man Has Good Reason To Feel Neglected." *Wall Street Journal* (October 18, 1993):A1

51. A. Fallon, "Culture in the Mirror: Sociocultural Determinants of Body Image," in *Body Images: Development, Deviance, and Change*, eds. T. F. Cash and T. Pruzinsky (New York: Guilford Press, 1990), p. 93. See: P. Rozin and A. Fallon, "Body Image, Attitudes to Weight and Misperceptions of Figure Preferences of the Opposite Sex: A Comparison of Men and Women in Two Generations," *Journal of Abnormal Psychology* 97:(1988):342–345.

52. M. Mishkind, J. Rodin, L. R. Silberstein, and R.H. Striegel-Moore, "The Embodiment of Masculinity: Cultural, Psychological, and Behavioral Dimensions," *American Behavioral Scientist* 29 (1986): 549.

53. M. Mishkind, J. Rodin, L. R. Silberstein, and R.H. Striegel-Moore, "The

Embodiment of Masculinity: Cultural, Psychological, and Behavioral Dimensions," *American Behavioral Scientist.* 29 (1986): 549.

54. See: M. B. King and G. Mezey, "Eating Behavior in Male Racing Jockeys," *Psychological Medicine* 17(1987):249–253; S. N. Steen, R. A. Oppliger, and K. D. Brownell, "Metabolic Effects of Repeated Weight Loss and Regain in Adolescent Wrestlers," *Journal of the American Medical Association* 260(1988):47–50.

55. M Millman, *Such a Pretty Face: Being Fat in America* (New York: Berkeley Books, 1980), p.225.

56. David Crawford. *Easing the Ache: Gay Men Recovering from Compulsive Disorders* (New York: Dutton, 1990), p. 126.

57. M. Mishkind, J. Rodin, L. R. Silberstein, and R. H. Striegel-Moore, "The Embodiment of Masculinity: Cultural, Psychological, and Behavioral Dimensions," *American Behavioral Scientist.* 29 (1986): 545–562.

58. A. D. Mickalide, "Sociocultural Factors Influencing Weight Among Males," in Arnold M. Andersen, *Males with Eating Disorders* (New York: Brunner/Mazel, 1990), pp. 30–39.

59. D. Garner and P. E. Garfinkel, "Socio-Cultural Factors in the Development of Anorexia Nervosa," *Psychological Medicine* 10 (1980): 647–656 and D. M. Garner, P. E. Garfinkel, D. Schwartz, and M. Thompson, "Cultural Expectations of Thinness in Women," *Psychological Reports* 47 (1980): 483–491; E. D. Rothblum, "Women and Weight: Fad and Fiction," *The Journal of Psychology* 124 (1990): 5–24.

60. M. P. Warren, and R. L. Vande Wiele, "Clinical and Metabolic Features of Anorexia Nervosa," *American Journal of Obstetrics and Gynecology* 117 (1973): 435–449; H. Bruch, "Anorexia Nervosa and Its Differential Diagnosis," *Journal of Nervous Mental Disease* 141(1966): 555–566; D. J. Jones, M. M. Fox, H. H. Babigan, and H. E. Hutton, "Epidemiology of Anorexia Nervosa in Monroe County, New York: 1960–1976," *Psychosomatic Medicine* 42 (1980):551–558.

61. G. A. German, "Aspects of Clinical Psychiatry, in Sub-Saharan Africa," *British Journal of Psychiatry* 121 (1972): 461–479; J. S. Neki, "Psychiatry in South East Asia," *British Journal of Psychiatry* 123 (1973): 257–269; B. Dolan, "Cross-Cultural Aspects of Anorexia Nervosa and Bulimia,"*International Journal of Eating Disorders* 10 (1991): 67–78; See also: A. Furnham and P. Baguma, "Cross-Cultural Differences in the Evaluation of Male and Female Body Shapes," *International Journal of Eating Disorders* 15 (1994):81–89.

62. E. D. Rothblum, "Women and Weight: Fad and Fiction," *The Journal of Psychology* 124 (1990):5. See also: P. S. Powers, *Obesity: The Regulation of Weight* (Baltimore, MD: Williams and Wilkins, 1980).

63. A. Furnham and P. Baguma, "Cross-cultural Differences in the Evaluation of Male and Female Body Shapes," *International Journal of Eating Disorders* 15 (1994):81–89. Sobal and Stunkard note: "Obesity may be a sign of health and wealth in developing societies, the opposite of its meaning in developed countries." They also note the importance of evolution: "Through the millennia, obesity was probably not a possibility for most people. Limited supplies of food characterized the lives of many of our ancestors and are present in many developing societies today" (pp. 266–267). See: J. Sobal and A.J. Stunkard, "Socioeconomic Status and Obesity: A Review of the Literature," *Psychological Bulletin* 105(1989): 260–275.

64. Emily Bradley Massara, *¡Qué Gordita! A Study of Weight Among Women in a Puerto Rican Community* (New York: AMS Press, 1989).

65. Emily Bradley Massara, ¡Qué Gordita! A Study of Weight Among Women in a Puerto Rican Community (New York: AMS Press, 1989), p. 19.

66. Emily Bradley Massara, ¡Qué Gordita! A Study of Weight Among Women in a Puerto Rican Community (New York: AMS Press, 1989), p. 12.

67. Emily Bradley Massara, ¡Qué Gordita! A Study of Weight Among Women in a Puerto Rican Community (New York: AMS Press, 1989), p. 171.

68. Emily Bradley Massara, ¡Qué Gordita! A Study of Weight Among Women in a Puerto Rican Community (New York: AMS Press, 1989), p. 293.

69. Emily Bradley Massara, ¡Qué Gordita! A Study of Weight Among Women in a Puerto Rican Community (New York: AMS Press, 1989), p. 141.

70. Emily Bradley Massara, ¡Qué Gordita! A Study of Weight Among Women in a Puerto Rican Community (New York: AMS Press, 1989), p. 145.

71. Emily Bradley Massara, ¡Qué Gordita! A Study of Weight Among Women in a Puerto Rican Community (New York: AMS Press, 1989), p. 161.

72. Emily Bradley Massara, ¡Qué Gordita! A Study of Weight Among Women in a Puerto Rican Community (New York: AMS Press, 1989), p. 161.

73. It is important to point out that the extent of the spread of the Cult of Thinness depends on a variety of factors within a given ethnic community. Within this community are the beginnings of intra-cultural variations in terms of the degree of assimilation. As the second generation of Puerto Ricans in this community come to gain upward mobility, their susceptibility to the Cult of Thinness may grow as well. As one researcher notes, "Among the African-American and Latina women I interviewed, the degree to which thinness was imposed on them as girls depended upon whether their families' class had changed, the families' geographical location, the schools the children attended, and nationality." See: B. Thompson,"Food, Bodies, and Growing Up Female: Childhood Lessons About Culture, Race, and Class," in Feminist Perspectives on Eating Disorders, eds. P. Fallon, M. A. Katzman, and S. C. Wooley (New York: Guilford Press, 1994), p. 371.

74. Emily Bradley Massara, ¡Qué Gordita! A Study of Weight Among Women in a Puerto Rican Community (New York: AMS Press, 1989), p. 145.

75. Regarding American society, see:J. E.Smith and J. Krejci, "Minorities Join the Majority: Eating Disturbances Among Hispanic and Native American Youth," International Journal of Eating Disorders 10 (1991):179–186; L. K. G. Hsu, "Are the Eating Disorders Becoming More Common in Blacks?," International Journal of Eating Disorders 6 (1987): 113–124. Regarding non-Western societies as a whole, see: M. Nasser, "Comparative Study of the Prevalence of Abnormal Eating Attitudes Among Arab Female Students of Both London and Cairo Universities," Psychological Medicine 16 (1986):621–625; J. Sobal and A. J. Stunkard, "Socioeconomic Status and Obesity: A Review of the Literature," Psychological Bulletin 105 (1989): 260–275; A. Furnham and B. Baguma, "Cross Cultural Differences in the Evaluation of Male and Female Body Shapes," International Journal of Eating Disorders 15(1994): 81–89.; T.Furukawa, "Weight Changes and Eating Attitudes of Japanese Adolescents Under Acculturative Stress: A International Prospective Study," International Journal of Eating Disorders 15(1994): 71–79; L. L. Osvold and G. R. Sodowsky, "Eating Disorders of White Ethnic American Racial and Ethnic Minority American and International Women," Journal of Multicultural Counseling and Development 21(1993):143–154.

76. See: L. K. George Hsu, "Are Eating Disorders Becoming More Common in Blacks?," International Journal of Eating Disorders 6 (1987): 122.

77. See: T. J. Silber, "Anorexia Nervosa in Blacks and Hispanics," *International Journal of Eating Disorders* 5 (1986):127.

78. See: C. S. W. Rand and J. M. Kaldau, "The Epidemiology of Obesity and Self-Defined Weight Problems in the General Population: Gender, Race, Age, and Social Class," *International Journal of Eating Disorders* 9 (1990): 329–343; V. G. Thomas and M. D. James, "Body-Image, Dieting Tendencies, and Sex Role Traits in Urban Black Women," *Sex Roles* 18 (1988): 523–529; C. E. Rucker III and T. F. Cash, "Body Images, Body-Size Perceptions, and Eating Behavior Among African-American and White College Women," *International Journal of Eating Disorders* 12(1992):291–299.

79. K. K. Abrams, L. Allen, and J. J. Gray, "Disordered Eating Attitudes and Behaviors, Psychological Adjustment, and Ethnic Identity: A Comparison of Black and White Female College Students," *International Journal of Eating Disorders*, 14 (1993): 49–57; B. Dolan, "Cross-Cultural Aspects of Anorexia Nervosa and Bulimia: A Review," *International Journal of Eating Disorders* 10 (1991):67–78.

80. E. White, "Unhealthy Appetites," *Essence* (September 1991):28. See also: C. S. W. Rand and J. M. Kaldau, "The Epidemiology of Obesity and Self-Defined Weight Problems in the General Population: Gender, Race, Age, and Social Class," *International Journal of Eating Disorders* 9 (1990): 329–343.

81. R. Bray, "Heavy Burden," *Essence* (January 1992): 54.

82. M. H. Styles, "Soul, Black Women and Food," in *A Woman's Conflict: The Special Relationship Between Women and Food*, ed. Jane Rachel Kaplan (New York: Prentice-Hall, 1980), pp. 161–162. The roots of plants are the primary ingredients of soul food—for example, yams, sweet potatoes, turnips, and "greens" such as collards.

83. M. H. Styles, "Soul, Black Women and Food," in *A Woman's Conflict: The Special Relationship Between Women and Food*, ed. J. R. Kaplan (New York: Prentice-Hall, 1980), p. 163.

84. M. H. Styles, "Soul, Black Women and Food," in *A Woman's Conflict: The Special Relationship Between Women and Food*, ed. J. R. Kaplan (New York: Prentice Hall, 1980) pp. 174–175.

85. J. A. Ladner, *Tomorrow's Tomorrow: The Black Woman* (New York: Doubleday, 1971).

86. See: K. K. Abrams, L. Allen, and J. J. Gray, "Disordered Eating Attitudes and Behaviors, Psychological Adjustment, and Ethnic Identity: A Comparison of Black and White Female College Students," *International Journal of Eating Disorders* 14 (1993): 49–57.

87. See: K. K. Abrams, L. Allen, and J. J. Gray, "Disordered Eating Attitudes and Behaviors, Psychological Adjustment, and Ethnic Identity: A Comparison of Black and White Female College Students," *International Journal of Eating Disorders* 14 (1993): 55. See also: J. J. Gray, K. Ford, and L. M. Kelly, "The Prevalence of Bulimia in a Black College Population," *International Journal of Eating Disorders* 6 (1987):733–740.

88. K. S. Buchanan, "Creating Beauty in Blackness," in *Consuming Passions: Feminist Approaches to Weight Preoccupation and Eating Disorders*, eds. C. Brown and K. Jasper (Toronto, Canada: Second Story Press, 1993) p. 37.

89. This impression is substantiated in a recent study by Abrams et al. See: K. K. Abrams, L. Allen, and J. J. Gray, "Disordered Eating Attitudes and Behaviors, Psychological Adjustment, and Ethnic Identity: A Comparison of Black and White Female College Students," *International Journal of Eating Disorders* 14 (1993):49–57.

90. K. S. Buchanan, "Creating Beauty in Blackness," in *Consuming Passions: Feminist Approaches to Weight Preoccupation and Eating Disorders*, eds. C. Brown and K. Jasper (Toronto, Canada: Second Story Press, 1993) p. 37.

91. P. H. Collins, *Black Feminist Thought: Knowledge, Consciousness and the Politics of Empowerment* (Boston: Unwin Hyman, 1990). Cited in K. S. Buchanan, "Creating Beauty in Blackness," in *Consuming Passions: Feminist Approaches to Weight Preoccupation and Eating Disorders*, eds. C. Brown and K. Jasper (Toronto, Canada: Second Story Press, 1993), p. 79.

Chapter 9

1. Celia Kitzinger, "Depoliticising the Personal: A Feminist Slogan in Feminist Therapy, *Women Studies International Forum* 16 (1993):487–496.

2. This phrase originated with Carol Hanisch. See Carol Hanisch, "The Personal Is Political," in *The Radical Therapist*, ed. J. Aget (New York: Ballantine, 1971).

3. Cynthia D. Schrager, "Questioning the Promise of Self-Help: A Reading of Women Who Love Too Much," *Feminist Studies* 19 (1993):188.

4. See Susan Kano, *Making Peace with Food: Freeing Yourself from the Diet/Weight Obsession* (New York: Harper & Row, rev. ed. 1989).

5. Rita Freedman, *Bodylove: Learning to Like Our Looks—and Ourselves* (New York: Harper & Row, 1988).

6. Marcia Germaine Hutchinson, *Transforming Body Image: Learning to Love the Body You Have* (Freedom, CA: The Crossing Press, 1985).

7. Wendy Simonds, *Women and Self-Help Culture: Reading Between the Lines* (New Brunswick, NJ: Rutgers University Press, 1992), p. 227.

8. Elayne Rapping, "Hooked on a Feeling," *The Nation* (March 5, 1990):317.

9. Gloria Steinem, *Revolution from Within: A Book of Self-Esteem* (Boston: Little Brown, 1992), p. 3.

10. Carol Gilligan, Nona P. Lyons, and Trudy J. Hanmer, eds., *Making Connections: The Relational Worlds of Adolescent Girls at Emma Willard School* (Cambridge, MA: Harvard University Press, 1990). See also: R. G. Simmons, D. A. Blyth, E. F. Van Cleave, and D. Bush, "Entry into Early Adolescence: The Impact of School Structure, Puberty and Early Dating on Self Esteem," *American Sociological Review* 44 (1979):948–967.

11. Catherine Steiner-Adair, "The Politics of Prevention," In *Feminist Perspectives on Eating Disorders*, eds. Patricia Fallon, Melanie A. Katzman, and Susan C. Wooley (New York : Guilford Press, 1994), p. 381.

12. E. Debold, M. Wilson, and I. Malavé, *Mother-Daughter Revolution: From Betrayal to Power* (New York: Addison-Wesley, 1993), p. 129.

13. bell hooks, *Yearning: Race, Gender and Cultural Politics* (Boston, MA: South End Press, 1990.), p. 219.

14. In an interview, bell hooks defined spiritual life as follows: "Simply, it has to do with the fundamental belief in divine spirit—in god and in love as a force that enables one to call forth one's godliness and spiritual power. . . . Spiritual life has much to do with self-realization, the coming into greater awareness not only of who we are but our relationship within community which is so profoundly political." See: bell hooks, *Yearning: Race, Gender and Cultural Politics* (Boston, MA.: South End Press, 1990), pp. 218–219.

15. See, for example, Johan Vanderlinden, Jan Norre and Walter Vander-eycken, *A Practical Guide to the Treatment of Bulimia Nervosa* (New York Brunner/Mazel, 1992); Kelly D. Brownell and John P. Foreyt, eds., *Handbook of Eating Disorders: Physiology, Psychology, and Treatment of Obesity, Anorexia and Bulimia* (New York: Basic Books, 1986); David M. Garner and Paul E. Garfinkel, eds., *Handbook of Psychotherapy for Anorexia Nervosa and Bulimia* (New York: Guilford Press 1985).

16. See: Patricia Fallon, Melanie A. Katzman and Susan C. Wooley, Eds., *Feminist Perspectives on Eating Disorders* (New York: Guilford Press, 1994)

17. Mary Bergner, Pam Remer, and Charles Whetsell, "Transforming Women's Body Image: A Feminist Counseling Approach," *Women and Therapy* 4 (1985): 25–38; Orland Wayne Wooley, Susan C. Wooley, and Sue R. Dyrenforth, "Obesity and Women II: A Neglected Feminist Topic," *Women's Studies International Quarterly* 2 (1979): 81–92.

18. Cynthia D. Schrager, "Questioning the Promise of Self Help: A Reading of *Women Who Love Too Much*," *Feminist Studies* 19 (1993): 189. Schrager refers to the important work of feminist theorist bell hooks who suggested this idea in her book, *Talking Back*. See: bell hooks, *Talking Back* (Boston: South End Press, 1989).

19. H.G. Lerner, "12 Stepping It: Women's Roads to Recovery," *Lilith* (Spring 1991):16.

20. Celia Kitzinger, "Depoliticizing the Personal: A Feminist Slogan in Feminist Therapy," *Women's Studies International Forum* 16 (1993):487–496.

21. Alison Bass, "Boycott Called on 'Anorexic' Ads," *Boston Globe* (April 25, 1994): 16.

22. Virginia L. Ernster, "Women, Smoking, Cigarette Advertising and Cancer," *Women and Therapy*, 6 (1987): 217–237.

23. Harrison G. Pope and James I. Hudson, *New Hope for Binge Eaters: Advances in the Understanding and Treatment of Bulimia* (New York: Harper & Row, 1984).

24. Alan B. Levy, Katherine N. Dixon, and Stephen L. Stern, "How Are Depression and Bulimia Related?," *American Journal of Psychiatry* 146 (1989):167.

25. Robert Chianese, "The Body Politic," *Utne Reader* (May/June 1992): 69.

26. See: Mary C. Franklin, "Eating Disorders a Topic for Girls," *Boston Globe* (May 8, 1994): 43.

27. *Teaching about Eating Disorders: Grades 7–12* (New York: Center for the Study of Anorexia and Bulimia, 1983). For a fuller discussion of the range of eating disorder prevention programs, see: C. M. Shisslak and M. Crago, "Toward a New Model for the Prevention of Eating Disorders," in P. Fallon, M. A. Katzman, and S. C. Wooley, eds., *Feminist Perspectives on Eating Disorders* (New York: Guilford Press, 1994), pp. 419–437.

28. This quote was taken from a recent ad for NAAFA that appeared in *Dimensions* magazine. The national address for NAAFA is: National Association to Advance Fat Acceptance (NAAFA), POB 188620, Sacramento, CA 95818.

29. Frigga Haug, ed., *Female Sexualization: A Collective Work of Memory* (London, Verso, 1987).

30. Frigga Haug, ed., *Female Sexualization: A Collective Work of Memory* (London, Verso, 1987), p.13.

31. Rosy Martin and Jo Spence, "New Portraits for Old: The Use of the Camera in Therapy," in Rosemary Betterton, ed., *Looking on: Images of Femininity in the Visual Arts and Media* (London: Pandora), p. 268.

32. Rosy Martin and Jo Spence, "New Portraits for Old: The Use of the Camera in Therapy," in Rosemary Betterton, ed., *Looking on : Images of Femininity in the Visual Arts and Media,* (London: Pandora), p. 268

33. Rosemary Betterton, ed., *Looking on: Images of Femininity in the Visual Arts and Media* (London: Pandora, 1987), p. 209.

References

Abrams, K. K., L. Allen, and J. J. Gray. 1993. "Disordered Eating Attitudes and Behaviors, Psychological Adjustments and Ethnic Identity: A Comparison of Black and White Female College Students." *International Journal of Eating Disorders* 14: 49–57.

Adams, G. R. 1977. "Physical Attractiveness Research." *Human Development* 20: 217–240.

Alexander, Suzanne. "Egged on by Moms, Many Teen-Agers Get Plastic Surgery." *Wall Street Journal*. September 24, 1990, p. 1.

Allon, Natalie. 1975. "Fat Is a Dirty Word: Fat as a Sociological and Social Problem." Pp. 244–247 in *Recent Advances in Obesity Research* 1, edited by A.N. Howard. London: Newman Publishing.

American Academy of Facial Plastic and Reconstructive Surgery. 1988. *The Face Book: The Pros and Cons of Facial Plastic and Reconstructive Surgery*. Washington DC: Acropolis Books.

American Association of University Women. 1991. *Shortchanging Girls, Shortchanging America*. Washington, DC: American Association of University Women.

American Family Physician. 1984. "Hypothalamic Set-Point System May Regulate Weight Loss." March, p. 269.

American Heritage Dictionary of the English Language. 1973. New York: American Heritage Publishing Company and Houghton Mifflin Company.

Anderson, A. E. 1984. "Anorexia Nervosa and Bulimia in Adolescent Males." *Pediatric Annals* 12: 901–4, 907.

Applefield, Catherine. "Keeping Up with All the Fondas." *Billboard*, November 16, 1991, p. 52.

Aquinas, Thomas. 1993. "Summa Theologiae." In *The Body Social: Symbolism, Self and Society*, ed. Anthony Synnot. New York: Routledge.

Aristotle. 1912. "De Generatione Animalium." In *The Works of Aristotle*, edited by J. A. Smith and W. D. Ross. London: Oxford.

Aristotle. 1921. "Politicia." In *The Works of Aristotle*, eds. J. A. Smith and W. D. Ross. London: Oxford.

Attie, Ilana and J. Brooks-Gunn. 1987. "Weight Concerns as Chronic Stressors in Women." Pp. 218–252 in *Gender and Stress*, edited by Rosalind Barnett, Lois Biener, and Grace Baruch. New York: The Free Press.

Attie, Ilana and J. Brooks-Gunn. 1992. "Developmental Issues in the Study of Eating Problems and Disorders." Pp. 35–58 in *The Etiology of Bulimia Nervosa: The Individual and Familial Context*, edited by J. H. Crowther, D. L. Tennenbaum, S. E. Hobfoll, and M. A. P. Stephens. London: Hemisphere Publishing Corporation.

Ayscough, Florence. 1937. *Chinese Women Yesterday and Today.* Boston: Houghton Mifflin Company.

Banner, Lois W. 1983. *American Beauty.* New York: Knopf.

Bar-Tal, Daniel and Leonard Saxe. 1976. "Physical Attractiveness and Its Relationship to Sex-Role Stereotyping." *Sex Roles* 2: 123–133.

Bart, Pauline B. 1975. "Emotional and Social Status of the Older Woman." In *No Longer Young: The Older Woman in America. Proceedings of the 26th Annual Conference on Aging,* edited by P.B. Bart. Ann Arbor: University of Michigan Institute of Gerontology.

Bartky, Sandra L. 1988. "Foucault, Femininity and the Modernization of Patriarchal Power." Pp. 61–88 in *Feminism and Foucault: Reflections on Resistance,* edited by Irene Diamond and Lee Quinby. Boston: Northeastern University Press.

Bashford, James W. 1916. *China: An Interpretation.* New York: Abingdon Press.

Bass, Alison. 1994. "Boycott Called on 'Anorexic' Ads." *The Boston Globe,* 25 April, pp. 1, 16.

Belasco, Warren J. 1984. " 'Lite' Economics: Less Food, More Profit." *Radical History Review* 28–30: 254–278.

Bemis, K. M. 1978. "Current Approaches to the Etiology and Treatment of Anorexia Nervosa." *Psychological Bulletin* 85: 593–617.

Bennett, William and Joel Gurin. 1982. *The Dieter's Dilemma: Eating Less and Weighing More.* New York: Basic Books.

Bergner, Mary, Pam Remer, and Charles Whetsell. 1985. "Transforming Women's Body Image: A Feminist Counseling Approach." *Women and Therapy* 4: 25–38.

Berkowitz, L., and A. Frodi. 1979. "Reactions to a Child's Mistakes as Affected by His/Her Looks." *Social Psychology Quarterly* 42: 420–425.

Berland, Theodore. 1990. "Rating the Weight-Loss Clinics: How They Differ—What They Cost." *Consumer Digest* (May/June): 65–69.

Berman, Ruth. 1989. "From Aristotle's Dualism to Materialist Dialectics: Feminist Transformation of Science and Society." Pp. 224–255 in *Gender/Body/Knowledge: Feminist Reconstruction of Being and Knowing,* edited by Alison M. Jaggar and Susan Bordo. New Brunswick: Rutgers University Press.

Berscheid, Ellen and Elaine Walster. 1972. "Beauty and the Beast." *Psychology Today,* October, pp. 42–46, 74.

Berscheid, Ellen, Elaine Walster, and G. Borhnstedt. 1973. "The Happy American Body: A Survey Report." *Psychology Today,* November, pp. 119–131.

Betterton, Rosemary, ed. 1987. *Looking on: Images of Femininity in the Visual Arts and Media.* London: Pandora.

Billowitz, Lisa. 1992. "Breast Implants: in the Aftermath of Corporate Greed." *Sojourner* August, pp. 12–15.

Blake, C. Fred. 1994. "Foot-binding in Neo-Confucian China and the Appropriation of Female Labor." *Signs: Journal of Women in Culture and Society* 19: 676–712.

Blumstein, Peter and Pepper W. Schwartz. 1983. *American Couples: Money, Work, Sex.* New York: William Morrow.

Bordo, Susan. 1987. *The Flight to Objectivity: Essays on Cartesianism and Culture.* New York: SUNY Press.

Bordo, Susan. 1988. "Anorexia Nervosa: Psychopathology as the Crystallization of Culture." Pp. 87–117 in *Feminism and Foucault: Reflections on Resistance,* edited by Irene Diamond and Lee Quinby. Boston, MA: Northeastern University Press.

Bordo, Susan. 1989. "The Body and the Reproduction of Femininity: A Feminist

Appropriation of Foucault." Pp. 13–33 in *Gender/Body/Knowledge: Feminist Reconstructions of Being and Knowing*, edited by Alison M. Jagger and Susan R. Bordo. New Brunswick: Rutgers University Press.

Bordo, Susan. 1993. *Unbearable Weight: Feminism, Western Culture and the Body*. Berkeley: University of California Press.

Boskind-Lodahl, Marlene. 1977. "The Definition and Treatment of Bulimarexia: The Gorging/Purging Syndrome of Young Women." Doctoral Dissertation, Cornell University, *Dissertation Abstracts International* 38, 7147A.

Boskind-Lodahl, Marlene and William C. White. 1978. "The Definition and Treatment of Bulimarexia in College Women—A Pilot Study." *Journal of the American College Health Association* 27, October: 84–97.

Brand, P. E., E. Rothblum, and L. J. Solomon. 1992. "A Comparison of Lesbians, Gay Men and Heterosexuals on Weight and Restrained Eating." *International Journal of Eating Disorders* 11: 253–259.

Bray, R. 1992. "Heavy Burden." *Essence*, January, p. 54.

Britton, A. G. 1988. "Thin Is Out, Fit Is In." *American Health*, July/August, pp. 66–71.

Brouwers, Mariette. 1988. "Depressive Thought Content Among Female College Students with Bulimia." *Journal of Counseling and Development* 66: 425–428.

Brownell, Kelly D. and John P. Foreyt, eds. 1986. *Handbook of Eating Disorders: Physiology, Psychology and Treatment of Obesity, Anorexia and Bulimia*. New York: Basic Books, Inc..

Bruch, Hilde. 1966. "Anorexia Nervosa and Its Differential Diagnosis." *Journal of Nervous Mental Disease* 141: 555–566.

Bruch, Hilde. 1973. *Eating Disorders: Obesity, Anorexia and the Person Within*. New York: Basic Books.

Brumberg, Joan. 1982. "Chlorotic Girls 1870—1920: A Historical Perspective on Female Adolescence." *Child Development* 3: 1468–1477.

Buchanan, K. S. 1993. "Creating Beauty in Blackness." Pp. 36–52 in *Consuming Passions: Feminist Approaches to Weight Preoccupation and Eating Disorders*, edited by C. Brown and K. Jasper. Toronto, Ontario: Second Story Press.

Button, E. J. and Whitehouse. A. 1981. "Subclinical Anorexia Nervosa." *Psychological Medicine* 11: 509–516.

Canter, R. J. and B. C. Meyerowitz. 1984. "Sex-role stereotypes: Self-reports of behavior." *Sex Roles* 10: 293–306.

Cash, Thomas F. 1990. "The Psychology of Physical Appearance: Aesthetics, Attributes, and Images." Pp. 51–79 in *Body Images: Development, Deviance, and Change*, edited by Thomas Cash and Thomas Pruzinsky. New York: Guilford Press.

Cash, Thomas F., Winstead, B. A. and Janda, L. H. 1986. "The Great American Shape-Up." *Psychology Today*, April, pp. 30–37.

Chapkis, Wendy. 1986. *Beauty Secrets: Women and the Politics of Appearance*. Boston: South End Press.

Chernin, Kim. 1981. *The Obsession: Reflections on the Tyranny of Slenderness*. New York: Harper & Row.

Chianese, Robert. 1992. "The Body Politic." *Utne Reader*, May/June, pp. 63–71.

Chodorow, Nancy. 1978. *The Reproduction of Mothering: Psychoanalysis and the Sociology of Gender*. Berkeley: University of California Press.

Chodrow, Nancy. 1989. *Feminism and Psychoanalytic Theory*. New Haven: Yale University Press.

Christian Science Monitor. 1992. "Losing Weight: A Profitable Business." October 8, p. 8.

Cohn, Lawrence D. and Adler, Nancy E. 1992. "Female and Male Perceptions of Ideal Body Shapes: Distorted Views Among Caucasian College Students." *Psychology of Women Quarterly* 16: 69–79.

Collins, Patricia Hill. 1990. *Black Feminist Thought: Knowledge, Consiousness and the Politics of Empowerment.* Boston: Unwin Hyman.

Cooley, Charles Horton. 1962 [1909]. *Social Organization.* New York: Schocken Books.

Corliss, R. 1982. "The New Ideal of Beauty." *Time,* August 30, pp. 72–73.

Crandall, Christian S. 1988. "Social Contagion of Binge Eating." *Journal of Personality and Social Psychology* 55: 588–598.

Crawford, David. 1990. *Easing the Ache: Gay Men Recovering from Compulsive Disorders.* New York: Dutton.

Crisp, A. H. 1965. "Some Aspects of the Evolution, Presentation and Follow-Up of Anorexia Nervosa." *Proceedings of the Royal Society of Medicine* 58: 814–820.

Currie, Dawn and Valerie Raoul. 1992. "The Anatomy of Gender: Dissecting Sexual Difference in the Body of Knowledge." Pp. 1–2 in *The Anatomy of Gender: Women's Struggle for the Body,* edited by Dawn Currie and Valerie Raoul. Ottawa, Canada: Carleton University Press.

Dagnoli, J., and J. Liesse. 1990. "Kraft, ConAgra Go Head-to-Head in Healthy Meals." *Advertising Age,* October 22, p. 59.

Davies, Mel. 1982. "Corsets and Conception: Fashion and Demographic Trends in the Nineteenth Century." *Comparative Studies in Sociology and History* 24: 611–641.

Davis, Fred. 1992. *Fashion Culture and Identity.* Chicago: University of Chicago Press.

Davis, Jenifer and Robert Oswalt. 1992. "Societal Influences on a Thinner Body Size in Children." *Perceptual and Motor Skills* 74: 697–698.

Debold, E., M. Wilson and I. Malavé. 1993. *Mother-Daughter Revolution: From Betrayal to Power.* New York: Addision-Wesley.

DeFrank, Thomas M. 1989. "Tales from the Diet Trenches." *Newsweek,* September 11, p. 58.

Deveny, Kathleen. 1993. "Light Foods Are Having Heavy Going." *Wall Street Journal,* Mar 4, p. B1.

Dion, Karen. 1972. "Physical Attractiveness and Evaluation of Children's Transgressions." *Journal of Personality and Social Psychology* 24: 207–213.

Dion, Karen, Ellen Berscheid, and Elaine Walster. 1972. "What Is Beautiful Is Good." *Journal of Personality and Social Psychology* 24: 285–290.

Dion, K. K. 1974. "Children's Physical Attractiveness and Sex as Determinants of Adult Punitiveness." *Developmental Psychology* 10: 772–778.

Dolan, Bridget. 1991. "Cross-Cultural Aspects of Anorexia Nervosa and Bulimia: A Review." *International Journal of Eating Disorders* 10: 67–78.

Dornbusch, Sanford M., J. Merrill Carlsmith, Paula Duke Duncan, Ruth T. Gross, John A. Martin, Philip L. Ritter, and Bryna Siegel-Gorelick. 1984. "Sexual Maturation, Social Class, and the Desire to be Thin Among Adolescent Females." *Developmental and Behavioral Pediatrics* 5: 308–314.

Douglas, Mary. [1970]1973. *Natural Symbols: Explorations in Cosmology.* 2nd ed. London: Barrie and Jenkins.

Drewnowski, A. and D. K. Yee. 1987. "Men and Body Image: Are Males Satisfied with Their Body Weight?" *Psychosomatic Medicine* 49: 626–634.

Dreyfus, Hubert L. and P. Rabinow. 1983. *Michel Foucault: Beyond Structuralism and Hermeneutics.* Chicago: University of Chicago Press.

Duffin, Lorna. 1978. "The Conspicious Consumptive: Woman as an Invalid." Pp. 26–56 in *The Nineteenth Century Woman: Her Cultural and Physical World*, edited by Lorna Duffin and Sara Delmont. London: Croon Helm.

Dull, Diana. 1989. "Before and Afters: Television's Treatment of the Boom in Cosmetic Surgery." In 84th Annual Meeting of the American Sociological Association in San Francisco, CA.

Dullea, Georgia. 1991. "Big Diet Doctor Is Watching You Reaching for that Nice Gooey Cake." *New York Times*, December 1, p. 65.

Durkheim, Emile. 1961. *The Elementary Forms of Religious Life*. Trans. Joseph Ward Swain. New York: Collier Books.

Dworkin, Andrea. 1974. *Woman Hating*. New York: Dutton.

Eberhard, Wolfram. 1966. "Introduction." In *Chinese Footbinding: The History of a Curious Erotic Custom*, edited by Howard S. Levy. New York: Walton Rawls Publisher.

Economist. 1992. "The Price of Beauty." January 1, pp. 25–26.

Economist. 1993. "The Fitness Industry—Snow Motion." March 27, pp. 71–72.

Edgley, Charles and Dennis Brissett. 1990. "Health Nazis and the Cult of the Perfect Body: Some Polemical Observations." *Symbolic Interaction* 13: 257–279.

Ehrenreich, Barbara and Dierdre English. 1979. *For Her Own Good: 150 Years of Expert Advice to Women*. Garden City, NJ: Anchor Books.

Eisenstein, Zillah. *The Female Body and the Law*. 1988. Berkeley: University of California Press.

Elder Jr., Glen. 1969. "Appearance and Education in Marriage Mobility." *American Sociological Review* 34: 519–533.

Elder Jr., G. H., T. V. Nguyen, and A. Caspi. 1985. "Linking Family Hardship to Children's Lives." *Child Development* 56: 361–375.

Epstein, Cynthia Fuchs. 1988. *Deceptive Distinctions: Sex, Gender and the Social Order*. New Haven: Yale University Press and New York: Russell Sage Foundation.

Ernster, Viginia L. 1987. "Women, Smoking, Cigarette Advertising and Cancer." *Women and Therapy* 6: 217–237.

Evans, E. D., J. Rutberg, C. Sather, and C. Turner. 1991. "Content Analysis of Contemporary Teen Magazines for Adolescent Females." *Youth and Society* 23: 99–120.

Ewen, Stuart. 1976. *Captains of Consciousness: Advertising and the Roots of the Consumer Culture*. New York: McGraw Hill.

Ewen, Stuart, and Elizabeth Ewen. 1982. *Channels of Desire: Mass Images and the Shaping of American Consciousness*. New York: McGraw-Hill.

"Exercise Video: Toned Up and Taking Off—Again." 1994. *Video Marketing Newsletter* 13,19:3.

Fallon, April. 1990. "Culture in the Mirror: Sociocultural Determinants of Body Image." Pp. 80–109 in *Body Image: Development, Deviance and Change*, edited by Thomas Cash and Thomas Pruzinsky. New York: The Guildford Press.

Fallon, April E. and Paul Rozin. 1985. "Sex Differences in Perceptions of Desirable Body Shape." *Journal of Abnormal Psychology* 94: 102–105.

Fallon, Patricia, Melanie A. Katzman and Susan C. Wooley, eds. 1994. *Feminist Perspectives on Eating Disorders*. New York: Guilford Press.

Featherstone, Mike. 1982. "The Body in Consumer Culture." *Theory, Culture and Society* 2: 18–33.

Featherstone, Mike, Mike Hepworth, and Brian S. Turner, eds. 1991. *The Body: Social Process and Cultural Theory*. Newbury Park, CA: Sage Publications.

Feldman, Gayle. 1989. "On the Road to Recovery with Prentice-Hall, Balantine, et al." *Publisher's Weekly*, November 3, pp. 52–53.

Feldman, W., E. Feldman, and J. T. Goodman. 1988. "Culture vs. Biology: Children's Attitudes Toward Thinness and Fatness." *Pediatrics* 81: 190–194.

Ferguson, M. 1983. *Forever Feminine: Women's Magazines and the Cult of Femininity*. London: Heinemann Educational Books.

Forse, Armour and G. L. Blackburn. 1989. "Morbid Obesity:Weighing the Treatment Options." Unpublished Paper. Nutrition/Metabolism Laboratory, Department of Surgery, New England Deaconess Hospital, Harvard Medical School, Boston, MA.

Foucault, Michel. 1977. *Discipline and Punish: The Birth of the Prison*. Translated by Alan Sheridan. New York: Pantheon Books.

Fox, Mary Frank and Sharlene Hesse-Biber. 1984. *Women at Work*. Palo Alto, CA: Mayfield.

Franklin, Mary. 1994. "Eating Disorders a Topic for Girls." *The Boston Globe*, May 8, pp. 43, 45.

Freedman, Rita. 1986. *Beauty Bound*. Lexington, MA: D.C. Heath and Company.

Freedman, Rita. 1988. *Bodylove: Learning to Like Our Looks—and Ourselves*. New York: Harper & Row.

Freeman, Jo. 1975. "How to Discriminate Against Women Without Really Trying." Pp. 217–232 in *Women: A Feminist Perspective*, 2nd ed., edited by Jo Freeman. Palo Alto, CA: Mayfield.

Friedan, Betty. 1963. *The Feminine Mystique*. New York: Norton.

Fugh-Berman, Adriane. 1994. "Traning Doctors to Care for Women." *Technology Review*, February/March, pp. 34–40.

Furakawa, T. 1994. "Weight Changes and Eating Attitudes of Japanese Adolescents Under Acculturative Stress: A Prospective Study." *International Journal of Eating Disorders* 15: 71–79.

Furnham, A. and P. Baguma. 1994. "Cross-Cultural Differences in the Evaluation of Male and Female Body Shapes." *International Journal of Eating Disorders* 15: 81–89.

Furst, Lilian R. and Perter W. Graham, eds. 1992. *Disorderly Eaters: Texts in Self-Empowerment*. University Park: The Pennsylvania State University Press.

Gabb, Annabella. 1989. "Heinz Meanz Brandz." *Management Today*, July, pp. 64–70.

Garfinkel, Paul E. 1981. "Some Recent Observations on the Pathogenesis of Anorexia Nervosa." *Canadian Journal of Psychiatry* 26: 218–223.

Garner, David and Paul Garfinkel. 1980. "Socio-Cultural Factors in the Development of Anorexia Nervosa." *Psychological Medicine* 10: 647–656.

Garner, David M, Marion P. Olmsted, and Paul E. Garfinkel. 1983. "Does Anorexia Nervosa Occur on a Continuum? Subgroup of Weight Preoccupied Women and Their Relationship to Anorexia Nervosa." *International Journal of Eating Disorders* 2: 11–20.

Garner, David M., Marion P. Olmsted, Yvonne Bohr and Paul E. Garfinkel. 1982. "The Eating Attitudes Test: Psychometric Features and Clinical Correlates." *Psychological Medicine* 12: 871–878.

Garner, David M., Paul E. Garfinkel, Donald Schwartz, and Michael Thompson. 1980. "Cultural Expectations of Thinness in Women." *Psychological Reports* 47: 483–491.

Garner, David M. and Paul E. Garfinkel. 1979. "The Eating Attitudes Test: An Index of Symptoms of Anorexia Nervosa." *Psychological Medicine* 9: 273–279.

Garner, David M. and Paul E. Garfinkel, eds. 1985. *Handbook of Psychotherapy for Anorexia Nervosa and Bulimia*. New York: Guilford Press.

German, G. A. 1972. "Aspects of Clinical Psychiatry in Sub-Saharan Africa." *British Journal of Psychiatry* 123: 461–479.

Gilligan, Carol, Nona P. Lyons and Trudy J. Hanmer, eds. 1990. *Making Connections: The Rational Worlds of Adolescent Girls at the Emma Willard School*. Cambridge, MA: Harvard University Press.

Giltenan, Edward. 1990. "Food, Drink & Tobacco." *Forbes*, January 8, pp. 172–174.

Glassner, Barry. 1988. *Bodies: Why We Look the Way We Do (and How We Feel About It)*. New York: G. P. Putnam.

Gordon, Ann, Mari Jo Buhle, and Nancy Schrom. 1971. "Women in American Society: A Historical Contribution." *Radical America* 5: 3–66.

Gordon, Richard. 1988. "A Sociocultural Interpretation of the Current Epidemic of Eating Disorders." Pp. 151–163 in *The Eating Disorders*, edited by B. J. Blinder, B. F. Chaiting and R. Goldstein. New York: PMA Publishing Corp.

Gordon, Richard. 1990. *Anorexia and Bulimia: Anatomy of a Social Epidemic*. Cambridge, MA: Basil Blackwell.

Gray, James J. and Kathryn Ford. 1985. "The Incidence of Bulimia in a College Sample." *The International Journal of Eating Disorders* 4: 201–210.

Gray, J. J., K. Ford, and L. M. Kelly. 1987. "The Prevalence of Bulimia in a Black College Population." *International Journal of Eating Disorders* 6: 733–740.

Gray, S. 1977. "Social Aspects of Body Image: Perceptions of Normality and Weight and Affect on College Unergraduates." *Perceptual and Motor Skills* 10: 503–516.

Greenblatt, Augusta. 1981. "Women in Medicine." *The Phi Beta Kappa Journal* 61, Fall: 10–11.

Greenhalgh, Susan. 1977. "Bound Feet, Hobbled Lives: Women in Old China." *Frontiers* 2: 17—21.

Gull, William. 1974. "Anorexia Nervosa (Apepsia Hysterica, Anorexia Hysterica)." *Transactions of the Clinical Society of London* 7: 22–28.

Halmi, Katherine, James Falk, and Estelle Schwartz. 1981. "Binge-Eating and Vomiting: A Survey of a College Population." *Psychological Medicine* 11: 697–706.

Halmi, Katherine A. 1974. "Anorexia Nervosa: Demographic and Clinical Features in 94 Cases." *Psychosomatic Medicine* 36: 18–25.

Halpin, Zuleyma Tang. 1989. "Scientific Objectivity and the Concept of 'The Other.' " *Women's Studies International Forum* 12: 285–294.

Hamburg, B. A. 1980. "Early Adolescence as a Life Stress." In *Coping and Health*, edited by Seymour and Ursin Levine Holger. New York: Plenum.

Hanish, Carol. 1971. "The Personal Is Political." In *The Radical Therapist*, ed. J. Aget. New York: Ballantine.

Hansen, J. and E. Reed. 1986. *Cosmetics, Fashions, and the Exploitation of Women*. New York: Pathfinder Press.

Hart, Kathleen J. and Thomas H. Ollendick. 1985. "Prevalence of Bulima in Working and University Women." *The American Journal of Psychiatry* 142: 851–854.

Hartmann, Heidi. 1976. "Capitalism, Patriarchy and Job Segregation by Sex." *Signs* 1: 137–169.

Hatfield, E. and S. Spreche. 1986. *Mirror, Mirror: The Importance of Looks in Everyday Life*. Albany: State University of New York Press.

Haug, Frigga, ed. 1987. *Female Sexualization: A Collective Work of Memory*. London: Verso.

Hawkins, Raymond and Pam Clement. 1980. "Development and Construct Validation of a Self-Report Measure of Binge Eating Tendencies." *Addictive Behaviors* 5: 219–226.

Hesse-Biber, Sharlene, Alan Clayton-Matthews, and John Downey. 1987. "The Differential Importance of Weight Among College Men and Women." *Genetic, Social and General Psychology Monographs* 113: 511–528.

Hesse-Biber, Sharlene. 1989. "Eating Patterns and Disorders in a College Population: Are Women's Eating Problems a New Phenomenon?" *Sex Roles* 20: 71—89.

Hesse-Biber, Sharlene. 1991. "Women, Weight and Eating Disorders: A Socio-Cultural and Political-Economic Analysis." *Women's Studies International Forum* 14: 173–191.

Hesse-Biber, Sharlene and Margaret Marino. 1991. "From High School to College: Changes in Women's Self-Concept and Its Relationship to Eating Problems." *The Journal of Psychology* 125: 199–216.

Hesse-Biber, Sharlene. 1992. "Report on a Panel Longitudinal Study of College Women's Eating Patterns and Disorders: Noncontinuum versus Continuum Measures." *Health Care for Women International* 13: 375—391.

Hightower, Jim. 1975. *Eat Your Heart Out: Food Profiteering in America*. New York: Crown Publishers, Inc..

Hirsch, Jules and Jerome L. Knittle. 1968. "Effect of Early Nutrition on the Development of Rat Epididymal Fat Pads." *Journal of Clinical Investigations* 47: 2091–2098.

Hirsch, Jules and Jerome L. Knittle. 1970. "Cellularity of Obese and Nonobese Human Adipose Tissue." *Federation Proceedings* 29: 1516–1521.

Hodge, Carole N., Linda A. Jackson, and Linda A. Sullivan. 1993. "The 'Freshman 15': Facts and Fantasies About Weight Gain in College Women." *Psychology of Women Quarterly* 17: 119–126.

Holland, Dorothy C. and Margaret A. Eisenhart. 1990. *Educated in Romance: Women, Achievement and College Culture*. Chicago: University of Chicago Press.

Holleran, P. R., J. Pascale, and J. Fraley. 1988. "Personality Correlates of College Age Bulimics." *Journal of Counseling and Development* 66: 378–381.

hooks, bell. 1989. *Talking Back*. Boston: South End Press.

hooks, bell. 1990. *Yearning: Race, Gender and Cultural Politics*. Boston: South End Press.

Horvath, T. 1979. "Correlates of Physical Beauty in Men and Women." *Sexual Behavior and Personality* 7: 145–151.

Horvath, T. 1981. "Physical Attractivenss: The Influence of Selected Torso Parameters." *Archives of Sexual Behavior* 10: 21–24.

Hovell, M. F., C. R. Mewhorn, Y. Randle, and J. S. Fowler-Johnson. 1985. "Risk of Excess Weight Gain in University Women: A Three Year Community Controlled Analysis." *Addictive Behaviors* 10: 15–28.

Hsu, L. K. George. 1987. "Are the Eating Disorders Becoming More Common in Blacks?" *International Journal of Eating Disorders* 6: 113–124.

Hsu, L. K. George. 1988. "Classification and Diagnosis of the Eating Disorders." Pp. 235–238 in *The Eating Disorders: Medical and Psychological Basis of Diagnosis and Treatment*, edited by B. J. Blinder, B. F. Chaitin and R. S. Goldstein. New York: PMA Publishing.

Hutchinson, Marcia Germaine. 1985. *Transforming Body Image: Learning to Love the Body You Have*. Freedom, CA: The Crossing Press.

Jackson, Linda. 1992. *Physical Appearance and Gender: Sociobiological and Sociocultural Perspectives*. Albany: State University of New York.

Jacobus, Mary, Evelyn Fox Keller and Sally Shuttleworth. 1990. *Body/Politics: Women and the Discourses of Science*. New York: Routledge.

Jay, Nancy. 1981. "Gender and Dichotomy." *Feminist Studies* 7: 37–56.

Jones, D. J., M. Fox, H. M. Babigan, and H. E. Hutton. 1980. "Epidemiology of Anorexia Nervosa in Monroe County, New York: 1960–1976." *Psychosomatic Medicine* 42: 551–558.

Jones, Margaret. 1990. "The Rage for Recovery." *Publisher's Weekly*, November 23, pp. 16–24.

Kaminer, Wendy, 1990. "Chances Are You're Codependent Too." *New York Times Book Review*. February 11, pp. 1, 26.

Kano, Susan. 1989. *Making Peace with Food: Freeing Yourself from the Diet/Weight Obsession*. Rev. ed. New York: Harper & Row.

Karlen, Neal. 1995. "Greetings from MINNESOBER!" *New York Times*, May 28, p. 32.

Katz, Stan and Aimee Liu. 1991. *The Codependency Conspiracy*. New York: Warner Books.

Katzman, M. A., S. A. Wolchik, and S. L. Braver. 1984. "The Prevalence of Frequent Binge Eating and Bulimia in a Nonclinical College Sample." *International Journal of Eating Disorders* 3: 53–62.

Keller, Evelyn Fox. 1978. "Gender and Science." *Psychoanalysis and Contemporary Thought: A Quarterly of Integrative and Interdisciplinary Studies* 1:409—433.

Kendall, R. E., D. J. Hall, A. Hailey, and H. M. Babigan. 1973. "The Epidemiology of Anorexia Nervosa." *Psychological Medicine* 3: 200—203.

Kimbrell, A. 1992. "Body Wars: Can the Human Body Survive the Age of Technology?" *Utne Reader*, May/June, pp. 52–64.

King, M. B. and G. Mezey. 1987. "Eating Behavior in Male Racing Jockeys." *Psychological Medicine* 17: 249–253.

Kitzinger, Celia. 1993. "Depoliticizing the Personal: A Feminist Slogan in Feminist Therapy." *Women Studies International Forum* 16: 487–496.

Kleinberg, S. 1980. *Alienated Affections: Being Gay in America*. New York: St. Martin's Press.

Kolata, Gina. 1993. "Accord on Implant Suit Brings Flood of Inquiries." *New York Times*, September 11, p. 7.

Kolata, Gina. 1994a. "3 Companies Near Landmark Accord on Breast Implant Lawsuits." *New York Times*, March 24, p. B10.

Kolata, Gina. 1994b."Details of Implant Settlement Announced by Federal Judge." *New York Times*, April 5, p. A1.

Kunzle, David. 1977. "Dress Reform as Antifeminism: A Response to Helene E. Roberts' " 'The Exquisite Slave: The Role of Clothes in the Making of the Victorian Woman.' " *Signs* 2:570—579.

Ladner, J. A. 1971. *Tomorrow's Tomorrow: The Black Woman*. New York: Doubleday.

Lasègue, 1873. Ernest-Charles. "On Hysterical Anorexia." *Medical Times and Gazette*, September 6, pp. 265–266; September 27, pp. 367–369.

Lawrence, M. L. 1987. *Fed Up and Hungry: Women, Oppression and Food*. New York: Peter Bedrick Books.

Lazarus, George. 1995. "Nestle Thaws Out Hearty Portions to Beef Up Its Frozen-Entree Menus." *Chicago Tribune*, May 11, p. 2.

Lehrer, Susan. 1987. *Origins of Protective Labor Legislation for Women: 1905–1925.* Albany: Sate University of New York Press.

Lerner, H. G. 1991. "Twelve Stepping It: Women's Roads to Recovery." *Lilith* Spring: 15–17.

Lerner, R. M. and S. A. Karabenick. 1974. "Physical Attractiveness, Body Attitudes and Self-Concept in Late Adolescents." *Journal of Youth and Adolescence* 3: 307–316.

Lerner, R. M., S. A. Karabenick, and J. L. Stuart. 1973. "Relations Among Physical Attractiveness, Body Attitudes and Self-Concept in Male and Female College Students." *Journal of Psychology* 85: 119–129.

Lerner, R. M. 1982. "Children and Adolescents as Producers of Their Development." *Development Review* 2: 342–370.

Levenkron, Steven. 1983. *Treating and Overcoming Anorexia Nervosa.* New York: Warner Books.

Levine, M. P. and L. Smolak. 1992. "Toward a Model of the Developmental Psychopathology of Eating Disorders: The Example of Early Adolescence." Pp. 59–80 in *The Etiology of Bulimia Nervosa: The Individual and Familial Context,* edited by J. Crowther, D. L. Tennenbaum, S. E. Hobfall and M.A.P. Stephens. London: Hemisphere Publishing Corporation.

Levy, Alan B., Katherine N. Dixon, and Stephen L. Stern. 1989. "How Are Depression and Bulimia Related?" *American Journal of Psychiatry* 146: 162–169.

Levy, Howard. 1966. *Chinese Footbinding: The History of a Curious Erotic Custom.* New York: Walton Rawls Publisher.

Liebman, B. F. 1987. "Fated to be Fat?" *Nutrition Action Health Letter* 14, January/February, pp. 4–5.

Liesse, Julie. 1992. "Healthy Choice Growing Pains: Why ConAgra Will Spend $200M on Energizing Its Megabrand." *Advertising Age,* August 24, p. 1.

The Lifestyle Market Analyst. 1993. Wilamette, IL: Standard Rate and Data.

Lifshitz, F., N. Moses, C. Cervantes, et al. 1987. "Nutritional Dwarfing in Adolescents." *Seminar in Adolescent Medicine* 3: 255–266.

Light, Donald. 1986. "Corporate Medicine for Profit." *Scientific American* , December, pp. 38–45.

Lloyd, G. 1984. *The Man of Reason: The Male and Female in Western Philosophy.* London: Methuen.

Lowe, Marion. 1982. "Social Bodies: The Interaction of Culture and Women's Biology." Pp. 91–116 in *Biological Woman-The Convenient Myth,* edited by R. Hubbard, M. S. Henifin, and B. Fried. Cambridge, MA: Schenkman Publishing Co.

Lurie, Alison. 1983. *The Language of Clothes.* New York: Vintage Books.

Lutz, Sandy. 1990. "Weight Loss Market's Profits Are Fading." *Modern Healthcare,* February 19, p. 50.

MacCannell, D. and J. F. MacCannell. 1987. "The Beauty System." Pp. 206–238 in *The Ideology of Conduct: Essays in Literature and the History of Sexuality,* edited by N. Armstrong and L. Tennenhouse. New York: Methuen.

Maccoby, Eleanor and Carol Nagy Jacklin. 1974. *The Psychology of Sex Differences.* Stanford, CA: Stanford University Press.

Mariette, Mlle. Pauline. 1866. *L'Art de la Toilette.* Paris: Librairie Centrale.

Marino, D. D. and J. C. King. 1980. "Nutritional Concerns During Adolescence." *Pediatric Clinics of North America* 27: 125–137.

Martin, Emily. 1987. *The Women in the Body: A Cultural Analysis of Reproduction.* Boston: Beacon Press.

Martin, Rosy and Ho Spence. 1987. "New Portraits for Old: The Use of the Camera in Therapy." Pp. 267–279 in *Looking On: Images of Femininity in the Visual Arts and Media*, edited by Rosemary Betterton. London: Pandora, 1987.

Massara, Emily Bradley. 1989. *¡Qué Gordita! A Study of Weight Among Women in a Puerto Rican Community*. New York: AMS Press.

Mayer, Jean. 1973a. "Fat Babies Grow into Fat People." *Family Health* 5: 24–38.

Mayer, Jean. 1973b. "When to Start Dieting? At Birth." *Medical World News*. September, pp. 31–33.

Mazur, Allan. 1986. "U.S. Trends in Feminine Beauty and Overadaptation." *The Journal of Sex Research* 22: 281–303.

McCabe, V. 1988. "Facial Proportions, Perceived Age and Caregiving." Pp. 89–95 in *Social and Applied Aspects of Perceiving Faces*, edited by T. R. Alley. Hillsdale, NJ: Erlbaum.

McCanne, Lynn P. Fisher. 1985. "Correlates of Bulimia in College Students: Anxiety, Assertiveness, and Locus of Control." *Journal of College Student Personnel*, July, pp. 306–310.

McCarthy, M. 1990. "The Thin Ideal, Depression and Eating Disorders in Women." *Behavior Research and Therapy* 28: 205–215.

McGough, Robert.1989. "Icing on the Cake." *Financial World*, 17 Oct., pp. 22–24.

Mead, George Herbert. 1934. *Mind, Self, and Society*. Chicago: University of Chicago Press.

Mellin, Laurel M., Sarah Scully, and Charles E. Irwin. 1986. "Disordered Eating Characteristics in Preadolescent Girls." In *American Dietetic Assocation Annual Meeting*, Las Vegas, NV.

Merchant, Carolyn. 1989. *The Death of Nature: Women, Ecology, and the Scientific Revolution*. New York: Harper & Row.

Michie, Helena. 1987. *The Flesh Made Word: Female Figures and Women's Bodies*. New York: Oxford University Press.

Mickalide, A. D. 1990. "Sociocultural Factors Influencing Weight Among Males." Pp. 30–39 in *Males with Eating Disorders*, edited by A.M. Andusen. New York: Brunner Mazel.

Miller, Annetta, Karen Springen, Linda Buckley and Elisa Williams. 1989. "Diets Incorporated." *Newsweek*, September 11, pp. 56–60.

Miller, Cindee. 1992. "Convenience, Variety Spark Huge Demand for Home Fitness Equipment." *Marketing News*, March 16, p. 2.

Millman, Marcia. 1980. *Such a Pretty Face*. New York: Berkeley Books.

Mintz, L. B. and N. E. Betz. 1988. "Prevalence and Correlates of Eating Disordered Behaviors Among Undergraduate Women." *Journal of Counseling Psychology* 35: 463–471.

Mishkind, M. E., J. Rodin, L. R. Silberstein and R. H. Striegel-Moore. 1986. "The Embodimentof Masculinity: Cultural, Psychological, and Behavioral Dimensions." *American Behavioral Scientist* 29: 545–562.

Mitchell, Russell, Lois Therrien, and Gregory Miles. 1990. "ConAgra: Out of the Freezer." *Business Week*, June 25, pp. 24–25.

Moody's Industrial Manual 1993. New York: Moody's Investor Services, 1993.

Morgan, H. G. and G. F. M. Russel. 1975. "Value of Family Background and Clinical Features as Predictors of Long-Term Outcome in Anorexia Nervosa: Four Year Follow-Up Study of 41 Patients." *Psychological Medicine* 5: 355–371.

Morris, A., T. Cooper, and P. J. Cooper. 1989. "The Changing Shape of Female Fashion Models." *International Journal of Eating Disorders* 8: 593—596.

Moses, N., M. Banlilivy, and F. Lifshitz. 1989. "Fear of Obesity Among Adolescent Girls." *Pediatrics* 83: 393–398.

Nasser, M. 1986. "Comparative Study of the Prevalence of Abnormal Eating Attitudes Among Arab Female Studens of Both London and Cairo Universities." *Psychological Medicine* 16: 621–625.

Neki, J. S. 1973. "Psychiatry in South East Asia." *British Journal of Psychiatry* 123: 257–269.

Nemeroff, C. J., R. I. Stein, N. S. Diehl, and K. M. Smilack. 1994. "From the Cleavers to the Clintons: Role Choices and Body Orientation as Reflected in Magazine Article Content." *International Journal of Eating Disorders* 16: 167–176.

Newsweek. 1986. "You're So Vain." April 14, pp. 48–55.

Newsweek. 1987. "The Littlest Dieters." July 27, p. 48.

New York Times. 1926. "Weight Reduction Linked to the Mind." *New York Times*, February 24, p.6.

Nichter, Mark and Mimi Nichter. 1991. "Hype and Weight." *Medical Anthropology* 13: 249–284.

Nichter, Mimi and Nancy Vuckovic. 1994. "Fat Talk: Body Image Among Adolescent Girls." Pp. 109–131 in *Many Mirrors: Body Image and Social Relations*, edited by Nicole Sault. New Brunswick, NJ: Rutgers University Press.

Ogden, J. 1992. *Fat Chance! The Myth of Dieting Explained*. New York: Routledge.

Ondercin, Patricia. 1979. "Compulsive Eating in College Women." *Journal of College Student Personnel* 19: 153–157.

Orbach, Susie. 1978. *Fat Is a Feminist Issue*. NewYork: Berkeley Press.

Orbach, Susie. 1986. *Hunger Strike: The Anorectic's Struggle as a Metaphor of Our Age*. New York: W.W. Norton.

O'Reilly, Brian. 1989. "Diet Centers are Really in Fat City." *Fortune*, June 5, pp. 137–140.

Ortner, Sherry B. 1974. "Is Female to Male as Nature Is to Culture?" Pp. 67–87 in *Woman, Culture and Society*, edited by Michelle Zimbalist Rosaldo and Louise Lamphere. Stanford: Stanford University Press.

Osvold, L. L. and G. R. Sodowsky. 1993. "Eating Disorders of White Ethnic American, Racial and Ethnic Minority Americans and International Women." *Journal of Multicultural Counseling and Development* July 21: 143–154.

Pappas, Nancy. 1989. "Body by Liposuction." *Hippocrates: The Magazine of Health and Medicine* 3, May/June: 26–30.

Pateman, Carole. 1988. *The Sexual Contract* Cambridge: Polity Press.

Peele, Stanton. 1975. *Love and Addiction*. New York: New American Library.

Peele, Stanton. 1989. *Diseasing of America: Addiction Treatment Out of Control*. Lexington. MA: D. C. Heath and Co.

Pertschuk, M. M. Collins, J. Kreisberg, and S. S. Fager. 1986. "Psychiatric Symptoms Associated with Eating Disorders in a College Population." *International Journal of Eating Disorders* 5: 563–568.

Pohl, Aime. 1992. "Teen Magazines' Message to Girls: You Can Be Anything . . . Except Yourself." *Extra: A Publication of FAIR* (Fairness in Accuracy and Reporting) New York: FAIR/EXTRA!, p. 28.

Polivy, Janet and C. Peter Herman. 1985. "Dieting and Binging: A Causal Analysis." *American Psychologist* 40: 193–201.

Pollitt, Katha. 1982. "The Politically Correct Body." *Mother Jones Magazine*, May , p. 67.

Pope, Harrison G.. James I. Hudson. 1984. *New Hope for Binge Eaters: Advances in the Understanding and Treatment of Bulimia*. New York: Harper & Row.

Powers, P. S. 1980. *Obesity: The Regulation of Weight*. Baltimore, MD: Williams and Wilkins.

Project on the Status of Education of Women. 1982. *The Classroom Climate: A Chilly One for Women*. Washington, DC: American Association of Colleges.

Psychology Today. 1984. "Bulimia: Not for Women Only." March, p. 10.

Pyle, Richard, James Mitchell, Elke Eckert, Patricia Halvorson, Patricia Neuman, and G.M. Goff. 1983. "The Incidence of Bulimia in Freshman College Students." *International Journal of Eating Disorders* 2: 75–85.

Pyle, Richard L., Patricia A. Halvorson, Patricia A. Neuman, and James E. Mitchell. 1986. "The Increasing Prevalence of Bulimia in Freshman College Students." *International Journal of Eating Disorders* 5: 631–647.

Quincy, Matthew. 1991. *Diet Right!* Berkeley, CA: Conari Press.

Rand, C. S. W. and J. M. Kaldau. 1990. "The Epidemiology of Obesity and Self-defined Weight Problems in General Population: Gender, Race, Age, and Social Class." *International Journal of Eating Disorders* 9: 329–343.

Rapping, Elayne. 1990. "Hooked on a Feeling." *The Nation*, March 5, pp. 316–319.

Richardson, S. A., N. Goodman, A. H. Hastorf, and S. M. Dornbusch. 1961. "Cultural Uniformity in Reaction to Physical Disabilities." *American Sociological Review* 26: 241–247.

Roberts, Helene E. 1977. "The Exquisite Slave: The Role of Clothes in the Making of the Victorian Woman." *Signs: Journal of Women in Culture and Society* 2: 554–569.

Rodin, Judith, Ruth H. Striegel-Moore, and Lisa R. Silberstein. 1990. "Vulnerability and Resilience in the Age of Eating Disorders." Pp. 366–390 in *Risk and Protective Factors in the Development of Psychopathology*, edited by J. Rolf, Masten, et al. Cambridge, England: Cambridge University Press.

Rodin, Judith 1992. *Body Traps: Breaking the Binds That Keep You From Feeling Good About Your Body*. New York: William Morrow and Company.

Rodin, Judith, Lisa Silberstein, and Ruth Striegel-Moore. 1985. "Women and Weight: a Normative Discontent." Pp. 267–307 in *Psychology and Gender: Nebraska Symposium on Motivation*, edited by T.B. Sonderegger. Lincoln: University of Nebraska Press.

Rosaldo, Michelle Zimbalist and Louise Lamphere, eds. 1974. *Women, Culture, and Society*. Palo Alto,CA: Stanford University Press.

Rosaldo, Michelle Zimbalist. 1974. "Women, Culture and Society: A Theoretical Overview." Pp. 67–87 in *Women, Culture and Society*, edited by Michelle Zimbalist Rosaldo and Louise Lamphere. Palo Alto, CA: Stanford University Press.

Rosenkrantz, Paul, Susan Vogel, Helen Bee, and Donald Broverman. 1968. "Sex Role Stereotypes and Self-Concepts in College Students." *Journal of Consulting and Clinical Psychology* 32: 287–291.

Rosenthal, Elisabeth. 1992. "Commercial Diets Lack Proof of Their Long-Term Successs." *New York Times*, November 24, pp. A1, C11.

Rothblum, Esther D. 1990. "Women and Weight: Fad and Fiction." *The Journal of Psychology* 124: 5–24.

Rothblum, Esther D. 1992. "The Stigma of Women's Weight: Social and Economic Realities." *Feminism and Psychology* 2: 61–73.

Rozin, P. and A. Fallon. 1988. "Body Image, Attitudes to Weight and Misperceptions of Figure Preferences of the Opposite Sex: A Comparison of Men and Women in Two Generations." *Journal of Abnormal Psychology* 97: 342–345.

Rubin, Gayle. 1975. "The Traffic in Women." Pp. 157–210 in *Toward an Anthropology of Women*, edited by Rayna Reiter. New York: Monthly Review Press.

Rucker III, C. E. and T. F. Cash. 1992. "Body Images, Body-size Perceptions and Eating Behavior Among African-Americans and White College Women." *International Journal of Eating Disorders* 12: 291–299.

Ryan, William. 1971. *Blaming the Victim*. New York: Pantheon.

Ryle, J. A. 1939. "Discussions on Anorexia Nervosa." *Proceedings of the Royal Society of Medicine* 32: 735–737.

Sanz, Cynthia and Leah F. Mitchell. 1990. "Fitness Tycoon Jenny Craig Turns Weight Losses into Profit by Shaping Her Clients' Bottom Line." *People Weekly*, February 19, pp. 91–92.

Scanlon, Deralee. 1991. *Diets That Work*. Chicago: Contemporary Books.

Schaef, Ann Wilson. 1987. *When Society Becomes an Addict*. San Francisco: Harper & Row.

Schoenfielder, Lisa and Barb Wieser. 1983. *Shadow on a Tightrope: Writings by Women on Fat Oppression*. Iowa City: Aunt Lute Book Company.

Schrager, Cynthia D. 1993. "Questioning the Promise of Self-Help: A Reading of Women Who Love Too Much." *Feminist Studies* 19: 177–192.

Schrank, Jeffrey. 1977. *Snap, Crackle, and Popular Taste: The Illusion of Free Choice in America*. New York: Dell Publishing Co.

Schwartz, Donald M., Michael G. Thompson, and Craig L. Johnson. 1982. "Anorexia Nervosa and Bulimia: The Socio-Cultural Context." *International Journal of Eating Disorders* 1: 20–36.

Schwartz, Hillel. 1986. *Never Satisfied: A Cultural History of Diets, Fantasies and Fat*. New York: Free Press.

Segal, S. A. and C. B. Figley. 1985. "Bulimia: Estimate of Increase and Relationship to Shyness." *Journal of College Student Personnel* 26: 240–244.

Seid, Roberta Pollack. 1989. *Never Too Thin: Why Women Are at War with Their Bodies*. New York: Prentice Hall Press.

Sheldon, W.H. 1940. *The Varieties of Human Physique: An Introduction to Constitutional Psychology*. New York: Harper & Row.

Shenson, Douglas. 1985. "Will M.D. Mean More Dollars?" *The New York Times*, May 23, p. 27.

Shisslak, C. M. and M. Crago. 1994. "Toward a New Model for the Prevention of Eating Disorders." Pp. 419–437 in *Feminist Perspectives on Eating Disorders*, edited by P. Fallon, M. A. Katzman, and S. C. Wooley. New York: Guilford Press.

Shorter, Edward. 1987. "The First Great Increase in Anorexia Nervosa." *The Journal of Social History* 21: 69–96.

Silber, T. J. 1986. "Anorexia Nervosa in Blacks and Hispanics." *International Journal of Eating Disorders* 5: 121–128.

Silberstein, L. R., R. H. Striegel-Moore, C. Timko and J. Rodin. 1988. "Behavioral and Psychological Implications of Body Dissatisfaction: Do Men and Women Differ?" *Sex Roles* 19: 219–231.

Silverstein, Brett. 1984. *Fed Up! The Food Forces That Make You Fat, Sick and Poor*. Boston: South End Press.

Silverstein, Brett, Lauren Perdue, Barbara Peterson, Linda Vogel, and Deborah A.

Fantini. 1986. "Possible Causes of the Thin Standard of Bodily Attractiveness for Women." *International Journal of Eating Disorders* 5: 135–144.

Silverstein, Brett and Lauren Perdue. 1988. "The Relationship Between Role Concerns, Preferences for Slimness and Symptoms of Eating Problems Among College Women." *Sex Roles* 18: 101–106.

Simmons, Roberta and Florence Rosenberg. 1975. "Sex, Sex Roles, and Self-Image." *Journal of Youth and Adolescence* 4: 229–258.

Simmons, R. G., D. A. Blyth, E. F. Van Cleave, and D. Bush. 1979. "Entry into Early Adolescence: The Impact of School Structure, Puberty and Early Dating on Self-Esteem." *American Sociological Review* 44:948–967.

Simonds, Wendy. 1992. *Women and Self-Help Culture: Reading Between the Lines.* New Brunswich, NJ: Rutgers University Press.

Smith, Dinitia. 1984. "The New Puritans: Deprivation Chic." *New York Times Magazine*, June 11, pp. 24–29.

Smith, J. E. and J. Krejci. 1991. "Minorities Join the Majority: Eating Disturbances Among Hispanic and Native American Youth." *International Journal of Eating Disorders* 10: 179–186.

Smith, P. A. and E. Midlarsky. 1985. "Empirically Derived Conceptions of Femaleness and Maleness: A Current View." *Sex Roles* 12: 313–328.

Sobal, J. and A. J. Stunkard. 1989. "Socioeconomic Status and Obesity: A Review of the Literature." *Psychological Bulletin* 105: 260–275.

Sontag, Susan. 1972. "The Double Standard of Aging." *Saturday Review* 55, September 23, pp. 29–38.

Sours, J. A. 1969. "Anorexia Nervosa: Nosology Diagnosis, Developmental Patterns and Power-Control Dynamics." Pp. 185–212 in *Adolescence: Psychosocial Perspectives*, edited by Gerald Caplan and Serge Lebovici. New York: Basic Books.

Squire, Susan. 1983. *The Slender Balance: Causes and Cures for Bulimia, Anorexia, and the Weight-Loss/Weight-Gain Seesaw.* New York: G. P. Putnam.

Stake, J. and M. L. Lauer. 1986. "The Consequence of Being Overweight: A Controlled Study of Gender Differences." *Sex Roles* 17: 31–47.

Stanley, Alessandra. 1992. "A Softer Image for Hillary Clinton." *New York Times*, July 13, pp. B1, B4.

Stark, Rodney and W. S. Bainbridge. 1986. *The Future of Religion: Secularization, Revival and Cult Formation.* Berkeley: University of California Press.

Steen, S. N., R. A. Oppliger, and K. D. Brownell. 1988. "Metabolic Effects of Repeated Weight Loss and Regain in Adolescent Wrestlers." *Journal of the American Medical Association* 260: 47–50.

Steele, Valerie. 1985. *Fashion and Eroticism: Ideals of Feminine Beauty from the Victorian Era to the Jazz Age.* New York: Oxford University Press.

Stein, D. M., and P. Reichert. 1990. "Extreme Dieting Behaviors in Early Adolescence." *Journal of Early Adolescence* 10: 108–121.

Steinem, Gloria. 1992. *Revolution from Within: A Book of Self-Esteem.* Boston: Little Brown.

Steiner-Adair, Catherine. 1994. "The Politics of Prevention." Pp. 381–394 in *Feminist Perspectives on Eating Disorders*, edited by Patricia Fallon, Melanie Katzman, and Susan C. Wooley. New York: Guilford Press.

Stern, Gabriella. 1993. "The Anorexic Man Has Good Reason to Feel Neglected." *Wall Street Journal*, October 18, p. A1.

Stern, Gabriella. 1994. "Makers of Frozen Diet Entrees Start Some Diets of Their Own." *Wall Street Journal*, January 4, p. B10.

Stewart, Doug. 1989. "In the Cutthroat World of Toy Sales, Child's Play is Serious Business." *Smithsonian* 20, December, pp. 72–84.

Stone, Lawrence. 1977. *The Family, Sex and Marriage in England 1500–1800.* New York: Harper & Row.

Striegel-Moore, Ruth H. , Lisa R. Silberstein, and Judith Rodin. 1986. "Toward Understanding the Risk Factors for Bulimia." *American Psychologist* 41: 246–263.

Striegel-Moore, Ruth H. 1992. "Prevention of Bulimia Nervosa: Questions and Challenges." Pp. 203–223 in *The Etiology of Bulimia Nervosa: The Individual and Familial Context,* edited by Janis H. Crowther, Daniel L. Tennenbaum, Stevan E. Hobfoll, and Mary Ann Parris Stephens. Washington, DC: Hemisphere Publishing Corporation.

Striegel-Moore, Ruth H., Lisa R. Silberstein, Peter Frensch, and Judith Rodin. 1989. "A Prospective Study of Disordered Eating Among College Students." *International Journal of Eating Disorders* 8: 499–509.

Stunkard, J., E. E. d'Aquili, S. Fox and R. D. L. Filion. 1972. "Influence of Social Class on Obesity and Thinness in Children." *Journal of the American Medical Association* 221: 579–584.

Styles, M. H. 1980. "Soul, Black Women and Food." Pp. 161–176 in *A Woman's Conflict: The Special Relationship Between Women and Food,* edited by J. R. Kaplan. Englewood Cliffs, NJ: Prentice-Hall, Inc.

Swift, W. J., D. Andrews, and N. E. Barklage. 1986. "The Relation Between Affective Disorder and Eating Disorder: A Review of the Literature." *American Journal of Psychiatry* 143: 290–299.

Synnott, Anthony. 1993. *The Body Social: Symbolism, Self and Society.* New York: Routledge.

Tallen, Bette S. 1990. "Twelve Step Programs: A Lesbian Feminist Critique." *NWSA Journal* 2: 390–407.

Tamburrino, M., K. N. Franco, G. A. Bernal, B. Carroll, and A.J. McSweeney. 1987. "Eating Attitudes in College Students." *Journal of American Medical Women's Association* 42: 45–47, 50.

Taylor, Frederick. 1967. *Principles of Scientific Management.* New York: W.W. Norton and Co.

Teaching About Eating Disorders: Grades 7–12. Center for the Study of Anorexia and Bulimia, 1983.

Thelen, M. H., C. M. Lawrence, and A. L. Powell. 1992. "Body Image, Weight Control and Eating Disorders Among Children." Pp. 81–101 in *The Etiology of Bulima Nervosa: The Individual and Familial Context,* edited by J. H. Crowther, D. L. Tennenbaum, S. Hobfoll, , M.A.P. Stephens. Washington, DC: Hemisphere Publishing Corp.

Therrien, Lois. 1989. "The Food Companies Haven't Finished Eating." *Business Week*, January 9, p. 70.

Therrien, Lois. 1990a. "Beatrice Investors Will Just Have To Sit Tight." *Business Week*, March 12, p. 104.

Therrien, Lois. 1990b. "Kraft Is Looking for Fat Growth from Fat-Free Foods." *Business Week*, March 26, pp. 100–101.

Thomas, V. G., and M. D. James. 1988. "Body-Image, Dieting Tendencies and Sex Role Traits in Urban Black Women." *Sex Roles* 18: 523–529.

Thompson, Becky. 1994. "Food, Bodies, and Growing Up Female: Childhood Lessons About Culture, Race and Class." Pp. 355–378 in *Feminist Perspectives on Eating Disorders*, edited by P. Fallon, M. A. Katzman, and S. C. Wooley. New York: Guilford Press.

Thompson, M. G. and D. Schwartz. 1982. "Life Adjustment of Women with Anorexia Nervosa and Anorexic-Like Behavior." *International Journal of Eating Disorders* 1: 47–60.

Tiggermann, M. and E. D. Rothblum. 1988. "Gender Differences in Social Consequences of Perceived Overweight in the United States and Australia." *Sex Roles* 18: 75–86.

Tuana, Nancy. 1989. "The Weaker Seed: The Sexist Bias of Reproductive Theory." Pp. 147–71 in *Feminism and Science*, edited by Nancy Tuana. Bloomington: Indiana University Press.

Turner, Bryan S. 1984. *The Body and Society*. New York: Basil Blackwell.

U.S. Department of Commerce, Bureau of the Census. 1982. *Census of Service Industries*. U.S. Department of Commerce, Bureau of the Census, Washington, DC.

U.S. Department of Commerce. 1993. *U.S. Industrial Outlook 1993*. U.S. Department of Commerce, U.S. Government Printing Office, Washington, DC.

Vanderlinden, Johan, Jan Norre, and Walter Vandereycken. 1992. *A Practical Guide to the Treatment of Bulimia Nervosa*. New York: Banner/Mazel.

Vaughn, Brian and Judith Langlois. 1983. "Physical Attractiveness as a Correlate of Peer Status and Social Competence in Preschool Children." *Developmental Psychology* 19: 561–567.

Veblen, Thorstein. 1899 (original edition). *The Theory of the Leisure Class*. New York: Random House, The Modern Library.

Verrell, Gordon. 1995. "Lasorda's Loss." *The Sporting News*, March 6, p. 7.

Video Marketing Newsletter. 1992. "Exercise Video: Toned Up and Taking Off-Again." *Vogue*. 1957. "How To Look Like a Beauty." *Vogue*, Sept. 15, p. 157.

Walby, Sylvia. 1990. *Theorizing Patriarchy*. Cambridge, MA: Basil Blackwell, Ltd.

Walker, Chip. 1993. "Fat and Happy." *American Demographics*, January 1, pp. 52–57.

Ward, Adrienne. 1990. "Americans Step into a New Fitness Market." *Advertising Age*, December 3, pp. 33, 39.

Warren, M.P. 1983. "Physical and Biological Aspects of Puberty." Pp. 3–28 in *Girls at Puberty: Biological and Psychosocial Perspectives*, edited by J. Petersen and A. C. Brooks. New York: Plenum.

Warren, M. P. and R. L. VandeWiele. 1973. "Clinical and Metabolic Features of Anorexia Nervosa." *American Journal of Obstetrics and Gynecology* 117: 435–449.

Weber, Joseph. 1990. "The Diet Business Takes It on the Chins: It's Under Government Scrutiny for Hype and Misleading Ads." *Business Week*, April 16, pp. 86–88.

Weber, Max. 1947. *The Theory of Social and Economic Organization*. New York: Free Press.

Weber, Max. 1963. *The Sociology of Religion*. Trans. Ephraim F. Schoff. Boston: Beacon.

Weibel, Kathryn. 1977. *Mirror, Mirror: Images of Women in Popular Culture*. New York: Anchor Books.

Weight Watchers International, Inc. 1994. "Corporate Backgrounder."

White, E. 1991. "Unhealthy Appetites." *Essence*, September, pp. 28–29.

Wichmann, S. and D. R. Martin. 1992. "Exercise Excess: Treating Patients Addicted to Fitness." *The Physician and Sports Medicine* 20: 193–200.

Williams, Monte. 1990. "People to Watch." *Advertising Age*, December 3, p. 36.

Wilshire, Donna. 1989. "The Uses of Myth, Image, and the Female Body in Re-Visioning Knowledge." Pp. 92–114 in *Gender/Body/Knowledge: Feminist Reconstruction of Being and Knowing*, edited by Alison M. Jaggar and Susan Bordo. New Brunswick, NJ: Rutgers University Press.

Wiseman, Claire V., James J. Gray, James E. Mosimann, and Anthony H. Ahrens. 1992. "Cultural Expectations of Thinness in Women: An Update." *International Journal of Eating Disorders* 1: 85—89.

Wohl, Stanley. 1984. *The Medical Industrial Complex*. New York: Harmony Books.

Wolfe, Leslie. 1991. *Women, Work and School: Occupational Segregation*. Westview Press.

Wonderlich, Stephen. 1992. "Relationship of Family and Personality Factors in Bulimia." Pp. 103–126 in *The Etiology of Bulimia Nervosa: the Individual and Familial Context*, edited by J.H. Crowther, D.L. Tennenbaum, S.E. Hobfoll, and M.A.P. Stephens. London: Hemisphere Publishing Corporation.

Woodman, Sue. 1994. "Losing Fat Permanently." *Fitness*, March/April, pp. 38–39.

Wooley, Orlando Wayne, Susan Wooley, Sue R. Dyrenforth. 1979. "Obesity and Women II: A Neglected Feminist Topic." *Women's Studies International Quarterly* 2: 81–92.

Zuckerman, Diana M., Anne Colby, Norma C. Ware, and Judith S. Lazerson. 1986. "The Prevalence of Bulimia Among College Students." *American Journal of Public Health* 76: 1135–1137.

Index

Note: Individuals interviewed by the author are listed by the pseudonymous first name given in the text. Subjects concerning diet and weight-loss are listed under headings beginning with the word "diet." References to illustrations and figures are given in italics.

Peele, Stanton, 145n53
Peggy, 73
Personality, relationship to body type, 149n13
Pharmaceutical industry, emphasis on slim figure, 4
Phillip Morris, 38
Photographs
  influence on women's ideal body image, 124
  used to monitor body size, 71
Physical fitness, 45. *See also* Fitness industry
Plastic surgeon (from Boston), 96
Plastic surgery, 50–57, *52. See also* Cosmetic surgery
Playboy, changing portrayals of ideal body image, 4
Pre-teen girls, 96–102
Prevention magazine, 50
Principal, Victoria, 45
Pritikin Diet, 10. *See also* Diet plans
Puberty, causes increased body fat, 101. *See also* Body weight
Puerto Rican women, views of thinness, 108–9
Purging, 83, 85, 93. *See also* Bingeing; Bulimia; Laxatives

Quaker (food manufacturer), 38
¡Que Gordita! (Massara), 108

Ralston Purina, 37
Recovery books, 42–43, 113–14, 118
Recovery groups, 43, 118
Religious cults, 9, 10
Rene (graduate student), 64–67
Revolution from Within (Steinem), 114
Rita, 71
Rituals. *See also* Cult of Thinness
  calorie counting, 79
  cults, 10
  food watching, 77–79
  in French boarding schools, 76–77
  physical fitness, 45–46
  used to attain ideal body image, 12
RJR Tobacco, 38
Roberta, 70
Rodin, Judith, 104, 106
Rosenkrantz, Paul, 17
Rothblum, Esther, 108

Ruth, 72
Ryle, John A., 27

Sara, 93
Scales (bathroom), used to monitor body size, 71–72
Scarsdale Diet, 10, 41
Scarsdale Diet, 10. *See also* Diets
Schaef, Anne Wilson
  Meditations for Women Who Do Too Much, 43
  When Society Becomes an Addict, 145n53
Science, gender gap of pursuit, 19, 123, 132n11
Seid, Roberta, 25, 99, 105
Self-esteem, 114, 125–26
Self-help books, 42–43, 113–14, 118
Self-help groups, 118. *See also* Overeaters Anonymous; Take Off Pounds Sensibly
Self-image. *See also* Men's ideal body image; Women's ideal body image
  compared to actual body size, 100
  development, 58–59
  influenced by college, 93–94
Shana, 91
Sheldon, W. H., 60
Sherry Vitti diet, 78
Sivlerstein, Brett, views on food industry, 35
Skeleton (teenage fashion model), 120
Slim Fast, 144n48. *See also* Diet foods; Diet plans
Smithsonian, article on Barbie dolls, 28–29
Social activism, used to combat Cult of Thinness, 119–22
Social status, 13–14, 61, 90
Soul foods, 110. *See also* Foods
Spirituality, requirement for social change, 117
Steinem, Gloria, *Revolution from Within,* 114
Stephanie (college sophomore), 34–35
Stewart, Doug, 28–29
STJ clothing, 47
Stouffer's, 38
Stress, triggers eating disorders, 92, 131n4
Styles, Marva, 110
Success, differing definitions for men's and women's, 12. *See also* Mind/body dualism
Surgery, 50–57, *52. See also* Cosmetic surgery; Plastic surgery